SOLID-FUEL FURNACES & BOILERS

SOLID-FUEL FURNACES & BOILERS

by John W. Bartok, Jr.

GARDEN WAY PUBLISHING
CHARLOTTE, VERMONT 05445

Illustrations by Michael W. Jager
Frontispiece photo by William Koechling
Printed in the United States

Library of Congress Cataloging in Publication Data
Bartok, John W., 1936–
    Solid-fuel furnaces & boilers.

    Bibliography: p.
    Includes index.
    1. Furnaces. 2. Boilers. 3. Coal. Fuelwood.   I. Title. II. Title: Solid-fuel
    furnaces and boilers.
TH7140.B325   621.402'5   82–1003   ISBN 0–88266–264–3 AACR2

CONTENTS

ACKNOWLEDGEMENTS

This book was written to gather together information on the rapidly developing field of solid-fuel central heating and is based on information received from many sources. Many manufacturers and distributors have given freely of their time, expertise, and educational materials. Also many building officials and fire marshals through their questions have pointed out areas of concern that need further investigation.

I have been encouraged to assemble this material by my coworkers in the Cooperative Extension Service at the University of Connecticut, especially the energy Extension agents who receive calls on this subject daily.

I am indebted to my late father, John W. Bartok, Sr., and my mother, Mary F. Bartok, for instilling in me an appreciation for the value of solid fuels in our society and for teaching me how to fire the gravity wood/coal furnace in the old homestead.

My colleague, Professor Emeritus Edward L. Palmer, who has worked with me in developing educational materials and presenting over 150 programs on solid fuel heating systems, has offered many helpful suggestions.

I am especially appreciative of the efficient and accurate typing of the manuscript by Mary Ryan and Cheryl Hendrickson, the Agricultural Engineering Department secretaries.

Finally, I want to thank my family; my son, Philip, and daughter, Cynthia, for their help with the illustrations and the catalog, and my wife, Janet, who has given me support in writing this book.

PREFACE

A few years ago it was simple. You built a home and you installed an oil or gas furnace, or opted for electrical heating, whichever was least expensive in your area.

Then came the energy crisis of 1973, and all of this changed. Oil, gas, electricity, all became more expensive.

The hunt was on for alternatives that offered heat at lower costs. The first major move was back to the woodstove — Grandfather's or the more efficient foreign models, and next the efficient models manufactured in this country.

Then those who had installed woodstoves tired of the mess they made, hated the extra work they involved, longed for the good old days when the heating unit was down cellar and only occasionally thought about. At the same time more and more people, faced with ever-rising fuel bills, decided that *now* they had to change from their oil or gas furnaces, but they wanted something better than a stove. The trend was on toward central heating. At the beginning of the energy crisis, fewer than ten old-line manufacturers were making residential-size solid-fuel heating units. This number has now grown to more than 200.

Units More Complicated

These units, prospective buyers find, are more complicated than the old woodstove. The buyer faces a long list of questions that have to be answered intelligently if he is to make a wise choice when selecting a central heating system. As many prospective buyers came to me for those answers, the situation became obvious. They needed a book that would provide them with the information so their choice would be right for their situation.

The buyer needs to understand the developments in design. These include multifuel systems, long-term water storage units, and combustion chamber changes that increase efficiency to levels equal to gas- or oil-fired units.

Size Is Important

Another factor needing careful consideration is the size of the unit to install. We will discuss several methods for determining this. The demands for heat vary widely from a unit that will take the chill off the house in the early fall and late spring to the more commonplace need for a unit to keep the house cozy when the temperatures drop to well below zero.

Questions of safety are of greatest concern with solid fuels. This is why many states require that only units that have been listed by an approved testing agency be sold. Some states also require that a building permit be obtained and that the unit be installed by a licensed heating contractor. Because of the complexity of most installations and because you are dealing with the safety of your family, it makes sense to follow these regulations.

Convenience of operation is on the top of the list of questions of many buyers. It may be worth the extra $1,000 to $2,000 to get a self-feeding unit, particularly if no one is at home all day. Such a unit will free you of being a slave to your heating system, a slave who has to be around seven days a week. It will help you to avoid a heating unit that is like a newborn baby — crying to be fed every four hours, day and night.

In answering thousands of calls each year on heating system operating problems, I have concluded that a good understanding of how fuels burn is very important. Why did the coal fire die out? Questions like this make me realize a basic understanding of combustion and heat transfer is needed.

Air Quality

Finally, and this is something often overlooked, the buyer should be concerned about air quality, both inside and outside the home. Things like tight stovepipe connections and adequate chimney draft will keep your air smelling fresh. High firebox temperatures and dry fuel will reduce the outside visual pollution and the complaints of your neighbors. With proper operation of your heater you may actually be reducing the amount of harmful pollution from the fuel it takes to heat your home.

You will find that the subjects in this book are not difficult to understand, and even more important, that if you do understand them, you will avoid what could be expensive and dangerous errors as you purchase, install, and operate a solid-fuel central heating system.

1 AN OVERVIEW

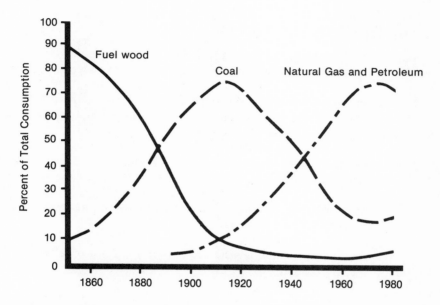

Fig. 1-1. U.S. fuel use patterns. (U.S. Bureau of Mines and Federal Energy Administration)

The decade of the 1980s is a transition period from one fuel age to another (fig. 1-1). It can be identified by rapidly escalating prices for oil and gas, and the introduction of many new alternate energy heating systems and concepts.

Similar transitions occurred in the 1880s when the world shifted from wood to coal, and in the 1930s when oil and gas became the predominant fuels.

Even back in 1744, Ben Franklin, in his *Account of the New-Invented Fireplaces,* concluded that, "Wood, our common fuel, which within these 100 years might be had at every man's door, must now be fetched near 100 miles to some towns and makes a very considerable article of expense of families."

In contrast to the former periods when the new fuel and its equipment were readily available, the new fuel for this transition period has not yet been identified. A combination of several energy sources may fill our needs over the next generation, possibly one or more of the following: solar, synthetic fuels, biomass, or nuclear. All have their proponents and opponents, their advantages and disadvantages. In the meantime we may have to rely on wood and coal, the older fuels, to get us through the transition period.

Your Options

You may be looking for alternatives to avoid the rapidly escalating heating prices that are eating into your budget. You have several options:

1. You can join the migration of people leaving the cold northern sections of the country and move to the South where only a little supplemental heat is needed.

2. You can build one of the new energy-efficient houses that reduce your fuel consumption to as little as one-tenth of your present usage.

3. You can go solar with either a passive or active system that provides up to 70 percent of your home's needs.

4. You can join the millions who elect to rely on America's natural resources of coal or wood.

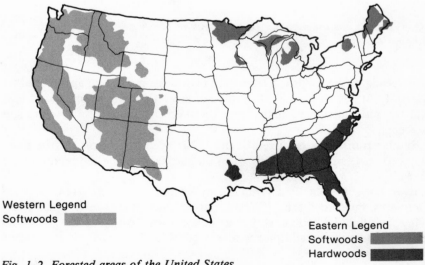

Western Legend
Softwoods

Eastern Legend
Softwoods
Hardwoods

Fig. 1-2. Forested areas of the United States

Both coal and wood are readily available in the colder areas of the country. Wood, a renewable, natural resource, is found on forest land that comprises more than 25 percent of the land area of the continental United States (fig. 1-2). In most areas our forests have been poorly managed over the past generation, and thinning, pruning, and clearing out the scrub trees can increase productivity and supply a significant portion of our heating needs for the next few years.

Wood is generally a good fuel choice if you live in rural areas where you are near the supply. As with any solid fuel, transportation costs can add significantly to the price. The farther you are from the source, the greater the cost.

Dry Hardwood Best

Hardwood (from trees having leaves) is the better choice as it contains almost twice as much heat per cord as softwood (trees with needles). The burning time for hardwoods is also longer, resulting in less stoking of the furnace. In either case your wood should be as dry as possible. A period of one to two years is needed to dry the wood from the 40 to 60 percent moisture content when first cut in the woods to the 20 percent air-dried content for wood that will burn well in the furnace. Let the sun remove the extra 100 to 250 gallons of moisture found in a cord, and you'll gain an extra 20 to 30 percent in the heat that isn't needed to boil off this water. You will also eliminate or diminish operating problems such as creosote buildup in the stovepipe and chimney.

With the large increase in the number of wood-fired heating units installed, a new fuel wood supply industry has developed. To date much of the wood has come from logging residues, land clearing for development, and forest improvement. Prices for wood have remained relatively stable.

The future will see greater use of densified wood and chips. Densified wood, which is logs, branches, and roots that have been chipped, then compressed into pellets or wafers, takes up less storage space than wood logs, and can be mechanically handled and fed into the heating unit.

There will also be greater use of other waste products such as manufacturing waste wood, compressed leaves, corn cobs, and straw. The availability of a fuel and its long-term supply outlook should be evaluated before you purchase a heating unit to burn it.

Hard and Soft Coal

COAL is another alternative, with large reserves in the United States (fig. 1-3). More than 6,000 mines are in operation. Some may be located within a few hundred miles of your home. The 1979 report of the U.S.

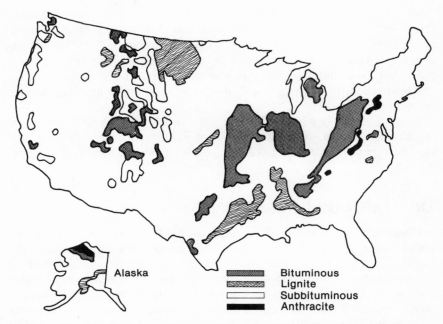

Fig. 1–3. Coal reserve areas of the United States. Adapted from Coal Facts, Natural Coal Assn., Washington, D.C.

Department of Interior, Bureau of Mines, showed the total identified reserves in the United States amount to over 1,600 billion tons. Of this over 475 billion tons are considered to be economic to mine today. Considering that the yearly production in 1980 was less than three-quarters of a billion tons, the reserves should last several hundred years.

ANTHRACITE (hard coal) is most commonly used for heating our homes. It is a long-burning, smokeless coal. Because the major supply area is eastern Pennsylvania, most of the Northeast uses this type of coal for heat. Some homeowners have trouble learning to burn anthracite because the ignition temperature is about 900° F., compared to 700° F. for bituminous coal and 550° F. for wood.

BITUMINOUS and subbituminous (soft coal) make up about 90 percent of our reserves. Mines can be found in at least twenty-six states.

Although the heat value (12,000 to 14,000 Btu/pound) is comparable with hard coal, soft coals generally create more pollution. These coals are dustier to handle and feel silky to the touch. Unlike wood, both hard and soft coal contain very little moisture.

Lignite is the lowest rank of coal. If you live in the Dakotas or the lower Mississippi Valley where most of the reserves are located, it may be worth considering as a fuel. Because it has a high moisture content and contains considerable amounts of mineral matter, it requires special furnaces and firing techniques.

With the rebirth of coal-fired central heating systems, a large number of manufacturers are adapting their units to be fed with stokers. Stokers automatically supply the fire with the proper amount of coal from a hopper or bin. This reduces some of the mess and time associated with tending the furnace.

Stove or Central Heat?

In evaluating solid-fuel heating units you will have to choose between stoves and central systems. This can be an easy decision if you already use a stove and want to move up to a unit that will heat the whole house.

Some central systems look like oversize stoves. A stove is a heating device that will heat part of your home with varying degrees of temperature from room to room. A stove can be located in any room where it can be attached to a chimney.

The central system is usually in the basement or furnace room, and pipes or ducts transfer the heat throughout the home. Most central systems are larger than stoves, and the firing rate is controlled automatically.

Here are some of the advantages of each type unit.

Stoves

1. Good for heating small areas from one to three rooms

2. Usually cost less to purchase and install

3. Will use less fuel as you are not heating the whole house

4. Available in ornate and attractive designs

5. Can be installed by the homeowner who is handy with basic tools.

Central System

1. Requires less attention, offers longer burning cycles with manual feed, occasional hopper refilling with automatic feed unit

2. More uniform temperatures throughout the house

3. Not located in expensive living space

4. Usually more efficient

5. Automatic controls make it safer

6. Keeps the ash and dirt in the basement

7. Available with oil/gas/electric backup in the same unit

8. Wood can be left in larger pieces; this means less splitting.

With the exception of the new water storage systems, all solid-fuel heating devices will cause a change in your life-style. With many, you will get up to a cool house many mornings during the winter. You can also plan on spending at least a half-hour a day to operate and maintain the furnace, not to mention the several weekends it will take to cut and prepare your wood supply if you opt for that fuel.

There is some security associated with solid fuels. If you go into the winter with a full woodshed or coal bin, oil or gas shortages and ice storms won't change the temperature of your house. It will still be an attractive place for your neighbors who were not so farsighted.

You will also be patriotic in helping to reduce the outflow of money from the United States by using our largest natural resources, coal or wood. Alternate fuel systems are most often installed by home owners who like home projects and who use their leisure time to improve their lives and enjoy the fruits of their labor.

Energy Conservation

Weatherization of your home is one of the most important steps you can take toward reducing your energy costs. It should be considered first, even before installing a stove or solid-fuel central system. Why? First, you will be saving some of our precious energy resources. Second, you will be reducing your own energy costs. Third, weatherization increases the value and salability of your home. Fourth, the money invested will pay higher returns than money placed in savings accounts, stocks, or bonds. Also the government allows a tax credit of 15 percent up to the limit of $2,000 for energy conservation costs.

About three-quarters of the homes in the United States are inadequately insulated. Unless you have recently built an energy-efficient home or have gone through an extensive weatherization project, your energy usage can be reduced by adding insulation or tightening up your home. Where do you begin?

Inspection

You can do this yourself or have it done through one of the audit programs. Inspections are being done through the utilities, heating contractors, and independent consultants at a reasonable price. The information is placed

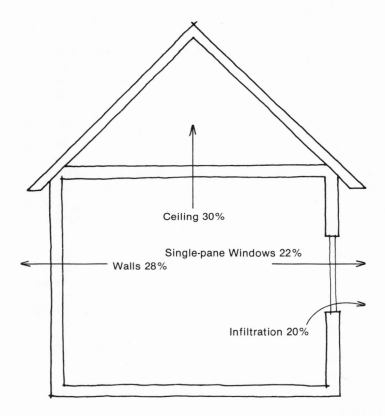

Fig. 1-4. In typical older house without insulation or storm windows, heat loss is greatest through ceilings.

in a computer for analysis. Usually the report you receive points out where you can get the best return for your money and what the payback period will be.

If you want to do the inspection yourself, there are several good books available including *In the Bank ... or Up the Chimney* available from Consumer Information Center, Pueblo, CO 81009, $1.70, and· *The Complete Energy Saving Home Improvement Guide,* Arco Publishing Co., 219 Park Ave. South, New York, NY 10003, $1.95.

Evaluation

Heat escapes from the house along many paths. The amount of heat that escapes by conduction over a particular period varies with the surface area through which it travels, how much insulation is present, and the difference in temperature between inside the house and outside. Adding insulation or reducing the inside temperature will help to conserve heat.

Our home also loses heat from infiltration. Each time the door is open some of the heated air escapes and an equal amount of cold air enters. Fuel is needed to warm this cold air. Infiltration also takes place through cracks around windows and doors, down poorly closed fireplace dampers, and under the sill of the house. Research has shown that the area of the cracks in an average home approximates a hole one foot square. As the wind speed increases, infiltration increases rapidly. That is why, on cold, windy days, it is hard to find a warm, comfortable spot in most homes.

An easy way to evaluate your home is to compare it against the established Department of Energy guidelines. Knowing where the largest heat losses occur, figure 1-4, will help to establish priorities. If you have an older home with many deficiencies, it's best to work on the major areas first.

Infiltration

A tight house will have less than one air change per hour. To get this requires storm windows and doors, caulking around window and door frames, weatherstripping on basement windows and doors, and tight seals around water, sewer, and electrical openings. You can check for air leaks on a windy day by moving the back of your hand around these openings. A windbreak of closely spaced evergreens on the windward side of the house will also help to reduce infiltration.

Attic Insulation

The attic area is usually the easiest to insulate. If it is used just for storage, the insulation goes in the floor. If the rooms are heated, the insulation should be placed between the rafters. Check the map, figure 1-5, for the amount recommended for your area. In some localities it is less expensive to have the insulation blown in place by a contractor than it is to purchase and install it yourself.

Wall Insulation

In houses with frame walls the cavity should be filled with insulation. This is easy to do in houses under construction or being remodeled. In many existing homes, especially those with some insulation in place, it is more difficult. Get the advice of a competent builder or insulating contractor. Fiberglass, cellulose, or other type of blown-in insulation may be used but works best if a vapor barrier has been placed on the room side. This can be a vinyl wallpaper or low-permeability paint.

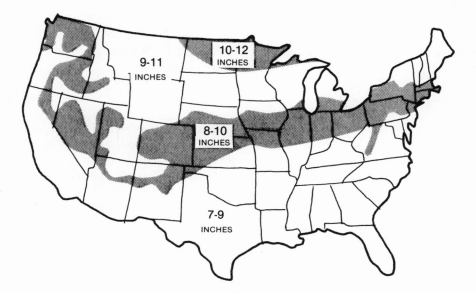

Fig. 1–5. Depth of ceiling insulation required can be estimated on this map.

Basement Walls or Ceiling

Which to insulate will depend on whether you heat the basement. If you are planning to install a central heating system in the basement, plan on insulating the walls. It will take about an extra cord of wood or ton of coal to heat your home each winter if the walls are not insulated. Heat loss through the walls, even though underground, can be substantial. Fiberglass and a stud wall can be used or you can apply board-type insulation covered with gypsum board. One to three inches of insulation should be used, with a vapor barrier on the basement side.

Heating System

Existing heating systems can be made more efficient by:

1. Having an annual cleaning and efficiency test

2. Insulating ducts and pipes in nonheated areas

3. Reducing the oil burner nozzle size after weatherization

4. Replacing the burner with a flame retention oil burner or high-efficiency gas burner with electronic ignition

5. Installing a flue damper to reduce heat loss up the chimney

6. Checking thermostat accuracy or replacing it with a night setback type.

Water Heater

The average home uses about fifteen kilowatt-hours of electricity, one gallon of oil, or 120 cubic feet of natural gas each day to heat the domestic hot water. Conservation where possible and the addition of insulation to systems having separate tanks will reduce costs significantly. Insulation kits are available for most separate heaters.

New Homes

Technology has advanced rapidly during the past few years to make new homes much more energy-efficient. In planning a new home purchase, read several books or spend time in the library researching latest methods of construction. Consider ideas such as:

1. Passive solar with an attached greenhouse

2. Berm or underground design

3. Multi-story

4. Water or rock thermal mass

5. Superinsulation or envelope construction.

As we progress through this decade of transition, energy may become more visible again. During the early period of wood and coal the homeowner was aware of how many logs or shovelfuls of coal it took each day to keep the house warm. He tried to conserve where possible so that he reduced his labor. As we went through the oil/gas era, energy became invisible and, because it was inexpensive, the homeowner seldom knew or gave a second thought to how much was being used.

As we enter the new era with the price of fuels escalating rapidly, we must once again get a feel for what energy is and how we can reduce its use. The switch to a solid-fuel central heating system will bring back this opportunity.

2 FUELS

Choosing a fuel is one of the first decisions you make when considering a new heating system. This may be easy if you own ten or more acres of woodland or you live near the entrance to a coal mine. You may even conclude that in your case it is best to burn a combination of wood and coal.

You will want to consider availability, convenience, and costs.

With oil, gas, or electric heat, you turn up the thermostat and the house warms up. The fuel is seldom given a second thought. We have been accustomed to automatic deliveries if we have oil or LPG equipment, or a continuous supply if we use natural gas or electricity. With wood or coal, getting the fuel is up to us. It may mean just a call to the local fuelwood or coal dealer or it may mean several weekends in the woods, cutting, splitting, and hauling wood home.

Both wood and coal are available in most of the cooler areas of the country. The Central States, New England, and the Middle Atlantic States have almost 50 percent of the total hardwood stands in the nation. Being a renewable resource and growing at the rate of up to one cord per acre per year, wood can supply the needs of many homes in these areas.

Coal also is in large supply, some of it in areas where wood is not as plentiful. In the southern tier of states, bituminous and subbituminous coal can be found, and the Dakotas have large supplies of lignite.

Alternate forms of wood other than stick wood are becoming more available, especially where extensive logging or manufacturing exists. Pellets, wafers, sawdust, and hogged fuel require less handling and lend themselves to bulk delivery and automatic feed systems. Many furnace and boiler manufacturers, seeing the advantages of bulk fuels, have integrated a stoker into their units.

The distance to the fuel supply greatly affects its cost. If you live near your own woodlot, hauling costs are limited to the cost of an old pickup or tractor and trailer and the few gallons of gas it takes each year. Hauling from a nearby state forest or mill requires a roadworthy vehicle with a cost per mile of twenty-five cents or more. Wood dealers who haul longer distances use trailer trucks with a cost per mile near one dollar. The accompanying table gives an estimate of cost added to the fuel you purchase just for the transportation, excluding labor for loading and unloading. As you can see, this adds a significant amount to the cost.

The same is true for coal. Delivery of a trailer-truck load from the mine in Pennsylvania to New Hampshire or Maine can cost twenty dollars or more per ton. If it is dumped in the dealer's coal yard and then hauled from there to your home, additional costs are added.

TABLE 2-1. TRANSPORTATION COSTS FOR SOLID FUELS INCLUDING DRIVER

Vehicle Type	Capacity-Cords	Stick Wood Cost/Mile-Cord	Cost/Cord 10 mi	50 mi	100 mi
Pickup ½–¾ ton	½–¾ ton	40–50¢	$4–5	na	na
Dump truck	1–2	20–40¢	$1–4	$10–20	na
Flatbed	3–5	10–20¢	$.20–.67	$5–10	$10–20
Trailer	6–10	6–10¢	na	$3–5	$6–10

Vehicle Type	Capacity-Tons	Coal and Bulk Wood Cost/Ton Cost/Mile-Ton	10 mi	50 mi	100 mi	500 mi
Pickup	½–1 ton	40–50¢	$4–5	na	na	na
Dump truck	1–3 tons	20–40¢	$2–4	$10–20	$20–40	na
Trailer	20–24 tons	2.5–5¢	na	na	$2.50–5	$12.50–25

Costs of Wood

Let's follow a cord of wood that you buy to see how it is handled and where the costs are. Bill Jacobs is a small fuelwood supplier in a rural area thirty miles outside of Central City. He hires workers to cut and haul the wood and operates a fuel yard on one of the main roads.

Not having any land himself, he contracts with farmers to clear land or purchases salvage rights to recently logged forests. The costs vary depending on his needs, the type of wood, and how accessible the land is for his equipment. They are usually about $5–20 per cord removed.

Bill has a two-man woods crew that works most of the year except on very rainy or snowy days. The equipment they use includes a four-cord flatbed truck, a jitterbug for the woods, a small crawler tractor and trailer when they get into rough land or when the snow gets deep, and a log splitter. Each man has two chainsaws.

Putting in long days, their daily output may be as much as eight cords when they have easy going such as clearing land, or it may drop to four cords when they are working on stony ground cutting tops left after a logging crew has moved out.

Before the wood reaches Bill's fuel yard it will have been handled once if it was loaded directly onto the truck, twice if it had to be hauled out of a wet or rocky area with a crawler. Another handling takes place when the wood is unloaded and stacked in the fuel yard to season for six to twelve months.

How to Save Money

If you buy the wood when it comes from the woodlot before cutting and seasoning, you will realize quite a savings. You are providing the labor for cutting and splitting, and the storage space. If you buy during the spring or summer, you may get an extra break on price and be assured that you have your winter's supply before the demand increases.

If you want your wood delivered seasoned, cut stove length, and split, Bill and his crew will do this for you. Cutting is usually done with a cord-wood saw attached to a tractor. It takes three men about a half-hour to cut a cord. Although most furnaces will take pieces ten inches or smaller in diameter, some of the larger pieces may have to be split, consuming another half-hour for two men.

Finally, the wood has to be delivered. Some dealers like Bill allow you to pick up the wood in their yard and haul it yourself. If you want Bill to deliver, this cost will have to be added. Table 2-2 gives you an idea of how much will be added.

A few companies have developed other methods for preparing the wood. Some use a skidder in the woods to pull tree-length pieces to a landing. Here they are cut into ten- to twenty-foot lengths and loaded onto a log truck with a cherry picker. These pieces can be taken directly to your yard or taken to an automated cut-off saw and splitting machine. The stove-length firewood is then carried by conveyor to a pile where it can be loaded with a bucket loader into the truck. This type of equipment is expensive, costing in

TABLE 2-2. VALUE ADDED TO A CORD OF FIREWOOD AT DIFFERENT STAGES*

In the woods	$10–20
Cut to four-foot lengths	$15–25
Delivered to fuel yard	$5–10
Cut to stove length	$10
Split	$5–10
Loaded and delivered to home (nearby)	$15
Seasoning	$10–15

*1982 Prices

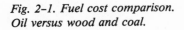

Fig. 2-1. Fuel cost comparison.
Oil versus wood and coal.

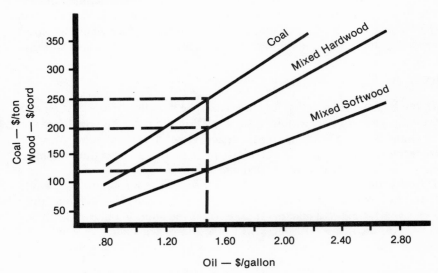

excess of $250,000, and a company using it must do a volume business to stay in operation.

Another method of handling the wood after it comes from the woods is to run it through a chipper or hogging machine. The pieces can be handled in bulk as chunks or sent through a compression press, which makes them into pellets or logs. You will pay for the energy needed to chop and compress the fuel but will gain handling ease.

Remember that anywhere along the path that you can short-circuit the fuel and provide some labor or storage, you will be reducing its price.

A similiar path can be developed for coal as it moves from the mines through the processing plant, the dealer, and then to you. Costs are added at each step of the way.

Now that you have an idea of the labor and costs involved in processing the wood or coal, what can you afford to pay? The simplest comparison is to look at the direct substitution cost of one fuel for another. Figures 2-1 through 2-3 do this for the most common fuels. The heating system efficiencies used are averages for the full season and not the maximum attainable under controlled conditions that many manufacturers use.

There are other costs to consider. The price of the furnace or boiler spread out over its useful life, usually twenty years, must be added to your total yearly heating costs. A new chimney, if you need one, and storage shed or bin for the fuel should also be considered. If you get your own wood the

Fig. 2-2. Fuel cost comparison.
Natural gas versus wood and coal.

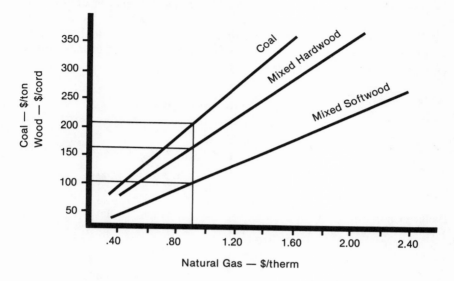

Fig. 2-3. Fuel cost comparison.
Electricity versus wood and coal.

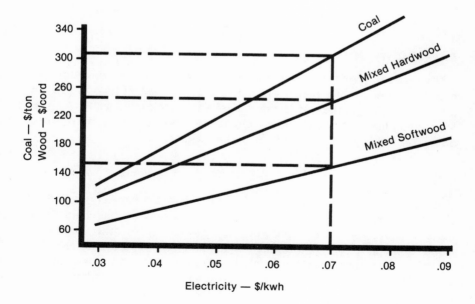

cost of the equipment should be depreciated over its useful life. Table 2-3 will help you calculate this for your own situation.

Once you have the total annual cost, divide by the number of cords of wood or tons of coal it takes to heat your home, and add it to your fuel cost. This is a true indication of your cost per cord or ton. Calculations for estimating the amount of solid fuel you will need per year can be made using the methods shown in Chapter 3.

TABLE 2-3. OTHER ANNUAL COSTS

| | Equipment Cost | | | Annual Cost | |
	Example — Yours	Life	Example — Yours
Furnace or boiler (installed)	$2,000	—	20	$100	—
New chimney	1,000	—	20	50	—
Chainsaw, maul, etc.	250	—	5	50	—
Hauling equip. for wood	1,000	—	10	100	—
Woodshed (10 × 12) for					
seven cords	500	—	20	25	—
Coal bin	100	—	10	10	—
Furnace repair and maint.,					
chimney cleaning, etc.				100	
Interest on investment					
10% of total investment				485	—
				$920	

Buying Wood

It's easier to purchase wood from a fuel yard or woodlot operator, but many homeowners get their own. The aroma of fresh-cut wood and the pleasure of being in the woods with nature are worth experiencing. Also, as long as you don't overdo it, the exercise from getting a cord or two of wood on a weekend is good for minds as well as bodies.

Sources

In the past few years I have seen many homeowners harvest most of the trees off their one- to two-acre house lots, leaving only the bare land. Once these trees are gone, the homeowners have to search for other sources of wood. Some develop a skill for scrounging along the roadsides after utility crews have gone through. Others search the landfills and dumps for discarded wood. These can be good sources, but you have to keep at it almost year-round to get enough to take you through the winter.

The salvage rights after logging crews have left are a good source in some areas. The wood in the tops amounts to one to two cords per thousand board feet of lumber removed. Sometimes you can get these for free, but more frequently you will have to pay $10 to $20 per cord. You may also be competing with the fuelwood dealer, especially on larger tracts. Before purchasing these rights, do a walking inspection of the land. Is it accessible with your equipment?

Areas for Cutting

Federal, state, and public utility land are sometimes available for selective cutting. Usually a permit is needed, and a small fee has to be paid. In some areas you can harvest only the trees that have been marked by the foresters. In others, you can harvest poor quality trees from a certain area. In some states these programs are so popular that there is a two- to three-year waiting list to get a permit.

If you live in a rural area where there are active farms, you may be able to work a deal with a local farmer to either clear some land he plans to put into fields or to keep his fencerows clean. You will probably be expected to dispose of the brush either by carrying it away or burning it.

Cutting the Wood

There is a lot of labor in handling wood. Each piece may be handled six to eight times between the woods and the furnace. Only experience will tell you whether it is better to cut it to four-foot lengths in the woods and later cut it to stove length, or to cut it to stove length and split it right in the woods. From my own experience, I think it would be better to handle it as four-foot lengths if you have a stove that takes very short pieces. With a furnace that will take pieces two feet or longer, you might as well cut it to length while it's attached to the tree. One advantage to a furnace over a stove is that it takes larger pieces. This reduces the amount of cutting and splitting. Most furnaces and boilers will take at least a twenty-four-inch length and ten-inch diameter, although the wood dries and burns better when split into smaller pieces.

Seasoning

Wood dries slowly. It often takes a year or more to reduce the moisture level to the approximate 20 percent air-dried level. In the drying process, most of the moisture and sap travels out through the ends of the pieces and not through the bark. You can speed drying by cutting the wood into shorter lengths, splitting it, and stacking it so the air currents will pass through the pile. Placing the wood in a solar dryer or on an asphalt driveway that absorbs the heat will speed the drying. Keep the wood off the ground so it doesn't absorb moisture.

The top of the pile should be covered so that the wood doesn't get wet every time it rains or snows. Use black plastic, roofing paper, metal roofing, or other material that will shed the water.

You may also want to build a woodshed. Rather than store the wood in your garage or basement and use up valuable space, you can locate a woodshed so that it is near the house. It will keep most of the bark and dirt outside.

The least expensive building is a pole building (fig. 2-4). The use of pressure-treated poles for long life, and rough framing lumber with a roof to match the home can make an attractive, functional building. A gravel base can be used for the floor. The building should be large enough to hold at least one winter's wood supply and preferably two. If you stack the wood six feet high, you need about twenty square feet of floor area for each cord stored. For example, a ten-foot-square building will hold five cords.

Heat Value

There is a lot of variation in wood both between different species and also between individual pieces. The amount of heat in a cord can vary from as much as twenty million Btu for a hardwood like hickory to less than ten million Btu for some of the soft woods (table 2-4). Even though there is this

TABLE 2-4

	Weight per Cord		Moisture Removed/Cord In Drying to	Approx. Heat Value
Species	Green	Air Dried	0 Moisture Content	60% eff. furnace
Ash	3840 lbs.	3440 lbs.	76 gal	16,500,000 BTU
Aspen	3440	2160	171	10,400,000
Beech	4320	3760	116	17,300,000
Yellow Birch	4560	3680	150	17,000,000
American Elm	4320	2960	186	14,300,000
Shagbark Hickory	5040	4240	131	20,400,000
Red Maple	4000	3200	153	14,000,000
Sugar Maple	4480	3680	132	17,400,000
Red Oak	5120	3680	220	16,900,000
White Oak	5040	3920	180	18,300,000
White Pine	2880	2080	126	9,500,000

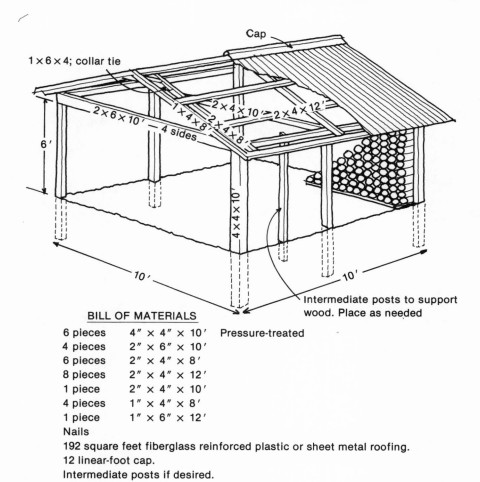

Cap

1 × 6 × 4; collar tie

2 × 6 × 10'
4 sides

1 × 4 × 8'
2 × 4 × 10'
2 × 4 × 8'
2 × 4 × 10'
2 × 4 × 12'

6'

4 × 4 × 10'

10'

10'

Intermediate posts to support wood. Place as needed

BILL OF MATERIALS

6 pieces	4″ × 4″ × 10′	Pressure-treated
4 pieces	2″ × 6″ × 10′	
6 pieces	2″ × 4″ × 8′	
8 pieces	2″ × 4″ × 12′	
1 piece	2″ × 4″ × 10′	
4 pieces	1″ × 4″ × 8′	
1 piece	1″ × 6″ × 12′	

Nails

192 square feet fiberglass reinforced plastic or sheet metal roofing.

12 linear-foot cap.

Intermediate posts if desired.

Fig. 2–4. Woodshed for five cords

wide variation, the heat value is constant on a weight basis. One pound of any dry wood except those containing resins, such as pine, has about 8,600 Btu. The resinous woods have a slightly higher heat value. In purchasing or getting your own wood, look for those having the highest heat values. You will not have to handle as much wood or fill the furnace as often if you burn them.

Dry wood is also worth more than green wood. As wood dries, moisture is removed and the volume shrinks. From the time the tree is cut to the time it burns in the firebox of the furnace, a cord of wood can lose 100 to 200 gallons of water (table 2-4). The shrinkage that takes place is from 2 to 4 percent for hardwoods and 3 to 6 percent for soft woods.

Fig. 2–6. This wood chipper, manufactured by the Morbark Industries, Inc., of Winn, MI, is capable of feeding itself, grinding up logs, and spewing the chips into a truck.

Other forms of wood and associated burners are available in some areas. These include forest residuals and manufacturing wastes such as tree tops, branches, foliage, bark, stumps, roots, sawdust, shavings, and wood scraps. In older processing plants over half of the tree ended up as waste. In today's modern plants this has been reduced to less than 10 percent.

Because these wastes vary so much in size and shape, it is difficult to use them directly for fuel in the furnace. To establish a uniform-sized piece, larger pieces are fed through a chipper or hogging machine (fig. 2-6). The rapidly rotating knives or blades cut the wood into chips. The size of the chip can be varied by the rate that the wood is fed in or the spacing of the knives. Chips for fuel are usually 1/2 to 1 inch long and 1/8 to 1/4 inch thick. Hogged fuel is usually larger, 1-1/2 inch long and 1/4 to 1/2 inch thick. It's important to have uniform pieces for automatic feeding systems and for good combustion. Chips may be screened to eliminate oversized or undersized pieces.

Pieces that are smaller than chip size (sawdust, shavings) may be burned directly in some furnaces, but some companies form these into compressed pellets, wafers, or logs. In this process, large pieces are screened out and the remainder dried and fed into a press operated by hydraulic pistons or rams.

The wood is compressed and forced out through dies and broken off into the length desired (table 2-5).

Chipped or compressed fuels are sold in bulk by the ton or unit (cubic unit = 100 cubic feet). You need a bin in the basement or outside for storage. In some cases a barn or other outbuilding is used. Unlike coal, which does not absorb moisture, the wood must be kept dry. Also, because the heat value is about half that of coal, a much larger area is needed if you want to store a winter's supply (table 2-6).

Because of the advantages of automatic stoking and reduced handling, alternate forms of wood will become more popular. This concept also lends itself to short rotation cropping of rapid-growth trees such as locust, cottonwood, sycamore, and alder. These trees can be planted at very close spacing on cleared forest land or poorer agricultural land and grown for five to fifteen years. At this time they would be harvested with equipment designed to cut and chip the tree. It is similiar to the forage harvesters used on dairy farms. The chips would then be dried and delivered to you in bulk. Research has shown that yields of up to ten tons per acre per year can be attained.

TABLE 2-5. PROPERTIES OF COMPRESSED WOOD

	Diameter	Length
Pellets	½–1 inch	½–1 inch
Wafers	2–4	¼–1
Pressed Logs	4	16

TABLE 2-6. ALTERNATE FORMS OF WOOD

Type	Weight-lbs/ft³	Volume-ft³/Ton	Heat Value-Btu/Ton
Sawdust			
green	10–13	150–200	8–10,000,000
kiln dry	8–10	200–250	14–18,000,000
Chips	10–30	67–200	16–20,000,000
Hogged	10–30	67–200	16–20,000,000
Pellets	40–50	40–50	18–20,000,000
Wafers	40–50	40–50	18–20,000,000
Pressed Logs			
wood	45	45	14–18,000,000
leaves	54	37	
Bark	10–20	100–200	16,–18,000,000
Rubber-pelletized	50–55	35–40	32–34,000,000

COAL

Coal is mined in twenty-six states, which makes it available within a few hundred miles of most of our homes. The eastern coal fields date back to the carboniferous era over 300 million years ago. Most of the coal in these fields is found in relatively thin seams well below the surface of the earth. It has a high heat value and high sulfur content. Most of the coal comes from mines as much as 600 feet below the surface.

The western fields are relatively new, dating back only 100 million years. With thicker seams nearer the surface, strip mining is practiced, using large power shovels. This coal has lower heat value and lower sulfur content. Much of it is used for generating electricity.

The formation of coal from the lush vegetative growth of the period alternating with earthquakes and volcanic eruptions that covered sections of the earth with water and sediment occurred over many years. This decaying matter was first formed into *peat,* a fibrous material burned as fuel in some parts of the world today. Peat is also commonly used as a fertilizer or soil amendment on farms and home gardens. Fuel peat must be dried and then compressed, and is sold as briquettes.

Lignite, a brown coal with a very high moisture content, is found mostly in the Dakotas and Texas. It has limited use as a home heating fuel because of its high moisture content and its tendency to crumble when dried and exposed to air. Its flames are long and smoky, and its heat value is low.

Furnaces and boilers that will burn lignite are available for those living close to the mines. It requires a strong draft and is fired with a thick bed. Storage is sometimes a problem. Spontaneous combustion can occur because of its low ignition temperature.

Subbituminous coal is harder than lignite, contains less moisture, and has a higher heat value. Mined mostly in Colorado, Washington, and Wyoming, it is used as a domestic fuel in those areas. Like lignite, it crumbles as it dries and can ignite spontaneously in piles or bins in which the temperature fluctuates and air circulates.

Studies by the U.S. Bureau of Mines indicate that subbituminous coal can be burned efficiently in hand-fired or stoker-fed furnaces. The rate of feed has to be higher to compensate for the lower heat value. A greater supply of secondary air is needed to burn the greater amount of volatile gases.

Bituminous coal (soft coal) is a black, soft, dusty fuel used in larger quantities for power generation and steel making. It is also widely used for home heating in areas near the mines. Large deposits are found in many states, but the most important beds are located in the Appalachian region.

Because it was formed under a variety of conditions, its physical characteristics vary widely. It is found as a high-volatile moist coal and a low volatile dry coal. It can have from 0.5 percent to over 8 percent sulfur content. When found as a caking coal (fuses when heated), it is used for manufacturing coke. The non-caking type is preferred for domestic use. Its heat value can be equal to or higher than anthracite because of its high volatile content.

Soft coal burns more easily than hard coal. The lower ignition temperature and the greater percentage of volatile matter make it easier to light and keep the fire going. The orange to red flames distinguish it from the blue flames of anthracite. Also the sulfur (rotten egg) odor and the greater amount of smoke and ashes make it less desirable.

Cannel coal is a dull, homogenous variety of bituminous coal sold almost exclusively for fireplace use. It ignites and burns easily with a candle-like flame and was used for lighting during the 1800s. Because of its origin in algae, fish, and other animal remains, it has a high volatile content and burns with a very hot flame. It is mined mostly in Kentucky, West Virginia, and Utah.

Anthracite (hard coal) is a smokeless, odorless, clean-burning fuel that is nearly pure carbon. The largest deposits are found in Pennsylvania, with smaller areas located in Virginia, Arkansas, Colorado, and Washington. In contrast to bituminous coal, which is often found near the surface and strip-mined, anthracite is located in deeper seams and is deep-mined. Hard coal is considered the best coal for home heating because it gives a long, even heat and the fire does not have to be tended as often. It is relatively expensive, mainly because of the higher mining and transportation costs, but is readily available from dealers in most of the colder sections of the country.

Coke is made by heating bituminous coal to drive off the volatile gases and other byproducts. It is almost pure carbon when it comes from the oven, and burns without smoke or fumes. Although not as readily available as a home heating fuel because of its demand in steel production, it can be found in some areas. It is easy to light and burns cleanly. Most furnaces that burn anthracite will burn coke, but because of its lighter weight and porous nature they must be refueled more often. The heat value is about the same as bituminous coal (12,500 to 14,500 Btu per pound but only 1,200–1,400 pounds can be obtained from a ton of coal).

Coal briquettes are made by compressing bituminous or lignite coal dust into small chunks. Often some type of binder or heat may be used to develop the necessary adhesion. They are sold by the bag and are not used much with domestic furnaces because they cost two to three times as much as regular coal.

Coal Sizes

Coal removed from a mine is sent to a processing plant. Here it goes through a crusher and then is sized through a series of screens. Impurities such as slate, bone, and sulfur-bearing material are removed. Some coal is washed to remove the dirt and then may be treated to reduce dustiness or to keep it from freezing when stored outside in winter.

Standard sizes have been adopted by the anthracite industry as an aid for consumers (table 2-7). Limits are also set on impurities. Some coal sold in the Northeast in the past few years has been taken from culm piles and it does not seem to meet these standards. Buyers have had trouble burning it. Purchase from a reputable dealer and you will get a quality coal.

TABLE 2-7. "STANDARD" ANTHRACITE SPECIFICATIONS

	Test Mesh-Round		Oversize	Undersize		Maximum Impurities		
Size of Coal	Through	Over	Max.	Max.	Min.	Slate	Bone	Ash
Broken	10"	3 1/4"		15%	7 1/2%	1 1/2%	2% OR	11%
Egg	3 1/4-3"	2 7/16"	5%	15%	7 1/2%	1 1/2%	2% OR	11%
Stove	2 7/16"	1 5/8"	7 1/2%	15%	7 1/2%	2%	3% OR	11%
Nut	1 5/8"	13/16"	7 1/2%	15%	7 1/2%	3%	4% OR	11%
Pea	13/16"	9/16"	10%	15%	7 1/2%	4%	5% OR	12%
Buckwheat	9/16"	5/16"	10%	15%	7 1/2%			13%
Rice	5/16"	3/16"	10%	17%	7 1/2%			13%
Barley	3/16"	3/32"	10%	20%	10%			15%
No. 4 Buck	3/32"	3/64"	20%	30%	10%			15%
No. 5 Buck	3/64"		30%	No Limit				16%

In the bituminous industry the screening and sizing practices vary widely. Table 2-8 shows those commonly used in stoves and furnaces.

TABLE 2-8. TEST MESH SIZE, BITUMINOUS

Size	Over	Through
Stoker	3/4"	1 1/4"
Nut	1 1/4	2
Egg	2	5

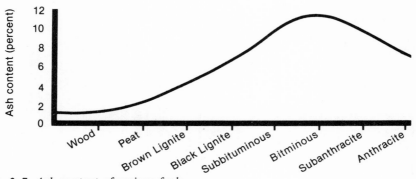

Fig. 2-7. Ash content of various fuels

Ash Content

Mineral matter and other impurities left after coal has burned are called ashes. Although coal may have an ash content as high as 30 percent when mined, less than 10 percent will remain in good coal as it leaves the processing plant (table 2-7). Remember that the ash is valueless mineral matter that you pay for at coal prices. Excess ash content also affects the operation of your furnace, requiring more frequent shaking and refueling. Get an indication of the quality of the coal you buy by asking the dealer for an analysis. It should tell you about the heat value, ash and sulfur content, and other physical properties.

Buying Coal

Coal is delivered to the coal dealer from the mine either by rail or truck. Although rail delivery is usually cheaper, some dealers do not have the facilities to handle a 100-ton carload at one time. Truck delivery is usually in twenty-ton loads. If you have a choice of dealers, it may pay to compare prices and quality.

The least expensive way to purchase coal is by the ton. Bagged coal, although cleaner and easier to handle, will cost $20 to $50 per ton extra. Bulk coal is usually delivered one to three tons at a time. A good access to within ten feet of the storage area is needed so trucks can deliver any time of the year.

Another popular method of buying coal is for coal-burning neighbors to contract for a trailer-truck load direct from the mine. It will save most of the profit that the coal dealer makes but it has some disadvantages. You may have to take delivery at one location or at least locations within a short distance of each other. The quantity that you receive may be difficult to

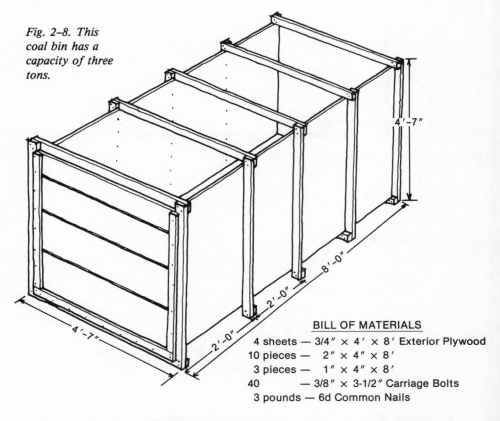

Fig. 2–8. This coal bin has a capacity of three tons.

4'-7"

8'-0"

2'-0"

2'-0"

4'-7"

2'-0"

BILL OF MATERIALS

4 sheets — 3/4" × 4' × 8' Exterior Plywood
10 pieces — 2" × 4" × 8'
3 pieces — 1" × 4" × 8'
40 — 3/8" × 3-1/2" Carriage Bolts
3 pounds — 6d Common Nails

measure, and the quality may not be as good unless you purchase from a reputable supplier. However, this is one way to save 15 to 30 percent on your winter's heating bill.

Storage

Although coal doesn't absorb moisture as wood does, it's best to store it under cover. This can be a bin in an unfinished basement, unused bulkhead, or oversized garage. It can be outside but should be near the house where it is convenient to get to in the winter. A coal delivery, particularly soft coal, will make a lot of dust, so consider this when you select a bin location.

The storage bin should be large enough to take delivery of several tons at one time. Using the measurement of forty cubic feet per ton, a four-by-four-by-eight-foot bin will hold about three tons (fig. 2-8).

When constructing a bin, consider ease of shoveling from it. If a bin is built in a basement with a concrete floor, shoveling will be much easier if coal spills onto the floor, and is shoveled up from there. The bin in fig. 2-8 can be built to work this way by omitting a bottom board on the front of the bin. In this way coal will not have to be lifted up out of the bin.

Combustion

Of the many calls and letters I get about problems with the operation of a particular unit, most reflect a poor understanding of the combustion process either by the manufacturer in the way the unit was designed, by the heating contractor in the way it was installed, or by the owner in how the unit is being operated.

Combustion is the chemical reaction that releases the heat in a fuel. This heat comes from the combination of carbon, hydrogen, and sulfur, the three major fuel elements, with oxygen. Light is also emitted from a solid fuel fire, and this can also be used to indicate how efficiently the furnace is operating.

Wood and coal burn in three stages that can be seen. These are moisture removal, volatile distillation, and combustion of fixed carbon.

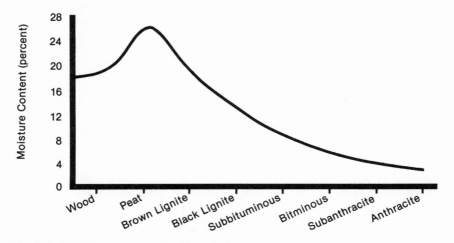

Fig. 2–9. Moisture content of various fuels

Moisture Removal

All solid fuels contain some moisture (fig. 2-9).

This moisture has to be heated, turned to steam, and removed before the fuel will burn. A fire is started with paper and finely cut pieces of dry kindling to develop a high enough temperature to dry out the fuel. Even wood that has been cut and dried under cover for a year or more still contains about 20 percent moisture by weight. If you watch a wood fire closely you can often see the steam and water being driven from the ends of the pieces of wood. Although bituminous and anthracite coal have very little inherent moisture, they may contain some surface moisture that has to be evaporated. Because of the high moisture level in lignite (20–50 percent), it is usually dried before it is used in the furnace.

Additional heat has to be supplied by burning paper and kindling to bring the fuel up to the ignition temperature before it will burn. The ignition temperature varies with the type of fuel, as can be seen in figure 2-10. Once the ignition temperature is reached, the fire will sustain itself. Because the ignition temperature is so high for anthracite coal, the fire will sometimes go out (cool to below the ignition temperature) if it is dampered down too much or if the firebox has a large surface area and you are operating a small fire.

Volatile Distillation

The next stage is to burn the volatile gases. For this to occur, additional heat and a higher temperature are needed. The volatile combustible gases

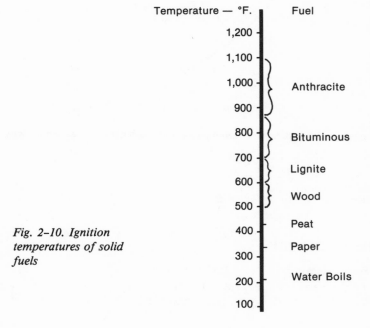

Fig. 2–10. Ignition temperatures of solid fuels

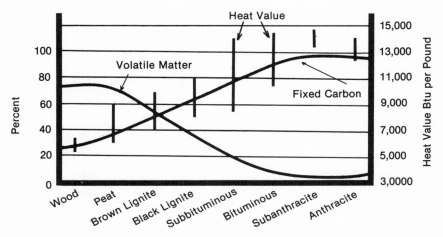

Fig. 2–11. Heat value and volatile matter in fuels. Read percentages at left for volatile matter and fixed carbon. Read Btu per pound at right for heat values.

are driven off, some at just above the ignition temperature and others as high as 1,200° F. The amount of volatile matter varies with the fuel (fig. 2-11). In wood it is very high and can amount to more than 50 percent of the total heat value. With anthracite being mostly solid carbon, the volatiles are usually less than 10 percent.

Space above the fire is needed to burn the volatiles. The greater the amount of volatiles, as in wood, the more space is needed for the mixing of the hot gases and the oxygen in the air for good combustion.

If adequate temperature is not maintained, as in a smouldering wood fire, the volatiles do not burn but are carried up the chimney as smoke. These mostly carbon particles can stick to the stovepipe or flue lining or collect in the pipe elbows as creosote. If there is excessive moisture present, as when burning green or wet wood, and the temperature of the flue gases drops below about 250° F., the liquid type of creosote that has the appearance of a sticky tar may drip out of the stovepipe at the joints. This can lead to a dangerous chimney fire condition. Because there are very little moisture and volatiles in coal, little creosote is formed. This is one advantage to burning coal.

Combustion of Fixed Carbon

After the volatiles are driven off, all that remains is the carbon or charcoal and a small amount of ash. This burns cleanly with very little visible smoke. Temperatures can reach 2,000° F. for wood and almost 2,700° F. for coal.

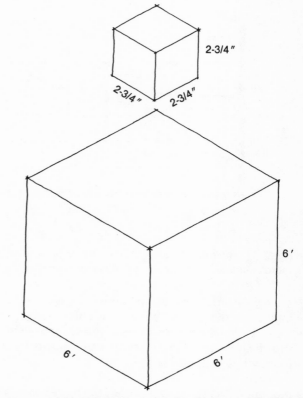

Fig. 2–12. To obtain complete combustion of a pound of coal, 200 cubic feet of air are needed.

2-3/4″

2-3/4″   2-3/4″

6′

6′   6′

The rate of combustion is controlled by the amount of air that is supplied to the fire. The more air, the faster the rate of burn and the more heat that is generated. Occasionally the answer to why a furnace is not supplying enough heat is that the drafts are not open enough or large enough. Table 2-9 shows the approximate air requirements for several solid fuels.

TABLE 2-9. APPROXIMATE THEORETICAL AIR REQUIREMENTS TO BURN

Fuel	Pounds per Pound of Fuel	Cubic Feet per Pound of Fuel
Wood	6.5	83
Peat	5.7	73
Lignite	6.2	79
Subbituminous	11.2	142
Bituminous	10.3	131
Anthracite	9.6	122
Coke	11.2	142

The rate of burn is also governed by the size of the fuel. This is why you use kindling wood to start a fire and large pieces for overnight fires. For coal the same is true.

If just the amount of air needed for combustion were fed to the fire, maximum efficiency would not be reached. This is because the mixing of the air and the combustibles is never complete and some unburnt particles would escape without coming into contact with the oxygen. An additional amount of air is needed to get more complete burning. This usually amounts to 20 to 50 percent (fig. 2-12).

On the other hand, excess air tends to cool the fire and takes more heat with it up the chimney. Remember that air is made up of 21 percent oxygen and 79 percent nitrogen. Only the oxygen is used for combustion. The nitrogen just robs the heat from the fire.

There is a point where the efficiency of combustion is the greatest. But another factor enters. This balance changes as the fire goes through different stages of the burn cycle. Greater amounts of the air are needed as secondary air above the fire when the volatiles are being burned (fig. 2-13). So you can see that the best we can hope for is an average efficiency that is as high as possible.

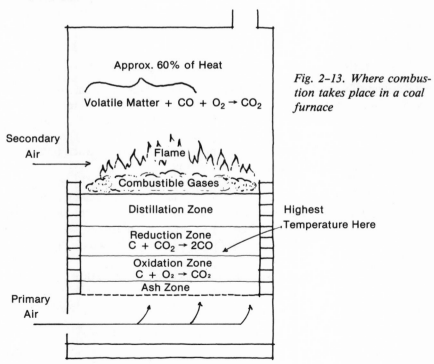

Approx. 60% of Heat

Volatile Matter + CO + $O_2 \rightarrow CO_2$

Secondary Air

Flame

Combustible Gases

Distillation Zone

Reduction Zone
$C + CO_2 \rightarrow 2CO$

Oxidation Zone
$C + O_2 \rightarrow CO_2$

Ash Zone

Primary Air

Highest
Temperature Here

Fig. 2–13. Where combustion takes place in a coal furnace

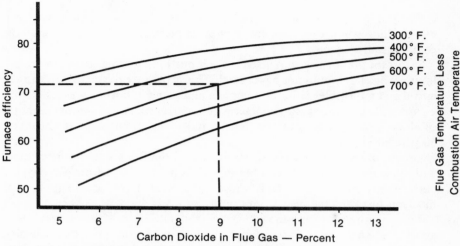

Fig. 2-14. Combustion efficiency for air-dried wood

We don't have quite the problems with combustion when we use fuel oil or gas. These are homogenous fuels, and once we set the fuel-air ratio on the furnace the efficiency remains almost constant except when the furnace starts.

In the combustion of solid fuels the following reaction takes place:

CARBON + OXYGEN = CARBON DIOXIDE + HEAT

Knowing this and using a carbon dioxide analyzer, a thermometer, and a calibration chart, the efficiency of a furnace can be determined. Manufacturers regularly test their units for efficiency under standard conditions to determine the effects of modifications they may make. Results of manufacturers' efficiency tests can give an indication of performance under ideal conditions and can be used to compare units.

You can get an indication of your own furnace efficiency by having a serviceman run the test for you. Most servicemen have efficiency testers. The test should be done after the fire has reached a steady state condition during the middle of the burn cycle. Because calibration charts for wood and coal are not normally found in test kits, they are included in two charts (fig. 2-14 and 2-15).

Average efficiencies for solid-fuel furnaces are given below:

Furnace and Fuel	*Average Efficiency %*
hand-fired, anthracite	60–75
hand-fired, bituminous	50–60
stoker-fired, bituminous	60–75

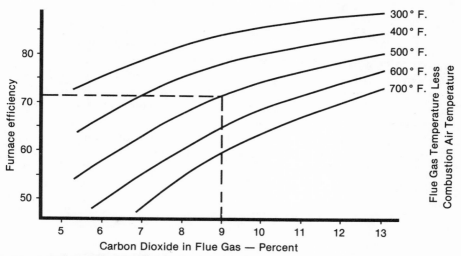

Fig. 2-15. Combustion efficiency for coal

Other factors that influence the combustion rate include:

1. The shape and volume of the combustion chamber. Too large a surface area cools the fire and reduces the amount of the volatile gases that are burned. If the surface area is too small, not enough heat is absorbed by the water or air heating the home and a higher stack temperature results. You lose about 1 percent in efficiency for each 25° F. that stack temperature is above 350° F.

2. The temperature of the combustion chamber should remain hot enough to get complete combustion. This is why firebrick or fireclay is commonly used. It maintains a higher, more uniform temperature in the burning zone.

3. The grate area and grate openings must be sized to support the fire and direct the air properly.

4. Secondary air above the fire determines the completeness of combustion. It is best supplied as close to the fuel bed as possible and in many small streams to give good mixing of the air with the combustible gases.

Gasification

Another burning method used mainly in larger commercial and industrial units has been applied to a home-size boiler by Essex Thermodynamics

Corp., Essex, Connecticut. The wood is baked in an oxygen-deficient atmosphere by a small fire in the bottom of the firebox. A highly flammable gas containing hydrogen and carbon monoxide is given off and is drawn through the bed of hot coals, mixed with oxygen, and then burned in a refractory chamber where temperatures are above 2,000° F. Nearly complete combustion is achieved, eliminating the smoke and creosote common with conventional updraft burning.

The heat is extracted by the boiler tubes and water jacket. Combustion efficiencies above 80 percent have been measured in home operation. The unit operates like the traditional oil- or gas-fired burner and cycles on only when heat is needed. The rest of the time the combustion process is dormant although the bed of coals remains hot.

ENVIRONMENTAL CONCERNS

Because solid fuels vary widely in their compostion and the combustion process also varies during the burn cycle, there is the potential for pollution of the air both inside and outside the home. Gases, vapors, and particulate matter are all emitted during burning. Some of these such as smoke are quite visible and can be used as an indication of incomplete combustion. Others such as carbon monoxide and hydrocarbons are toxic and in large enough quantities may affect your health.

SMOKE contains particles of carbon and other tarry substances. In large enough quantities and over a period of time it can stain the siding on your home. It can also injure plants and humans and is the partial cause of the smog in larger cities.

Wood and hard coal give off very little smoke if burned when dry with a good air supply. Soft coal and lignite, having more volatile matter, tend to produce greater amounts.

You can reduce the amount of smoke by:

1. Using a dry fuel

2. Loading with smaller charges, more often

3. Opening the draft for a few minutes after loading

4. Operating the furnace with a higher firebox temperature

5. Closing the stovepipe damper slightly to hold the gases in the firebox a little longer.

Smoke nuisance laws are on the books in some communities. These attempt to limit burning under certain weather conditions or pollution levels. They generally are hard to enforce.

ASHES are the solid matter left after the fuel is burned. The smaller particles may be carried out the chimney as part of the smoke. Larger particles settle to the bottom of the fire and are shaken through the grate into the ash pan.

Wood ashes contain potassium and calcium and are useful as a garden soil conditioner and fertilizer. You can apply about 10 pounds per 100 square feet per year. Each cord burned will yield 60-100 pounds.

Coal ashes vary widely in their content depending on what impurities were mixed in as the coal was formed. In general the ashes are high in silicon, aluminum, and iron, with traces of some of the heavy metals. Expect 200-300 pounds from a ton of coal. They are best used on icy walks in the winter or as a binder for gravel driveways.

SULFUR amounts in solid fuels very widely (table 2-10).

TABLE 2-10. SULFUR CONTENT OF FUELS

Fuel	Range of Amount — percent
Wood	.01–.05
Peat	.1–1.5
Lignite	.2–7
Subbituminous	.3–2
Bituminous	.5–8
Anthracite	.5–1.2
Coke	.5–1.5

In coals sulfur occurs as iron sulfide (pyrite), organic sulfur, and sulfates. Upon burning, most of this is changed to sulfur dioxide, a colorless gas with a pungent, irritating odor. Because it is highly soluble in water, it combines with the water vapor given off during combustion and forms a weak acid solution. This is harmful in two ways: it deteriorates the stovepipe and chimney flue lining, and it can, when exhausted into the atmosphere, combine with the moisture to form acid rain. This makes the soil and lakes more acid, which affects plant and animal life.

The problem can be minimized by using low sulfur fuels, an Environmental Protection Agency requirement in most states.

CARBON MONOXIDE is a colorless, almost odorless, tasteless, and non-irritating gas produced by the incomplete combustion of fuels. When breathed, it is absorbed through the lungs and gets into the blood stream. Because it has an affinity more than 200 times greater than oxygen for the hemoglobin in the blood, the oxygen is displaced and serious problems can occur (table 2-11).

Although nationally, automobiles generate much greater amounts of carbon monoxide (over 100 times as much), there have been some injuries and deaths with solid-fuel home heating units, mostly with coal. The amount of carbon monoxide formed in the furnace is very small if adequate secondary air is available above the fire, good mixing of the gases occurs, and a hot temperature is maintained in the firebox. It is also important to have a tight furnace, tight stovepipe joints, and adequate make-up air in well-insulated houses. A carbon monoxide detector is a good investment for safety and peace of mind if you are burning coal. Because carbon monoxide is lighter than air, the detector should be located near the ceiling in the area of the furnace. One unit that I have found to be effective is the Gas Sniffer manufactured by Revco Products, 441 East Columbine Ave., Santa Ana, CA 92707.

CARBON DIOXIDE (CO_2), one of the byproducts of combustion, is not generally considered an air pollutant because it is an essential part of plant and animal life cycles. Plants remove CO_2 from the air and through photosynthesis convert it to plant tissue. During the process, oxygen is put back into the air.

The concern comes because the level of CO_2 in the atmosphere is increasing at a rapid rate, mainly due to the large increase in the use of fossil fuels. This increase in the CO_2 level may result in a warming of the earth through the greenhouse effect although factors such as an increase in particulate matter or an increase in the cloud cover could offset this. One consoling fact is that manmade CO_2 production is less than 1 percent of the total each year. Most of it comes from biological degradation and the oceans.

Catalytic burners may be one way to reduce the amount of pollutants put into the air from wood and coal furnaces. They are like the catalytic converters in cars. The flue gases are brought into contact with a catalyst that lowers the temperature at which the creosote and soot will burn. The catalyst is noble metal, usually platinum, that has been coated onto a screen or honeycomb. This is then placed within the stovepipe.

These devices, although effective in converting up to 90 percent of the pollutants, have a few problems. The screen must be washed if it becomes coated with soot, and over a period of time ages so that it has to be replaced. It also may lose its effectiveness if you have a chimney fire. Another disadvantage is that it adds up to $100 to the cost of the furnace. Only a few manufacturers are offering these burners for sale.

In summarizing the environmental impact of solid fuel use, you may refer to table 2-12. All fuels create pollution, some more than others. Recent studies by the North Carolina State University School of Forestry Research concluded that wood fuel is a little cleaner than coal but has about the same net environmental effect as residual fuels such as fuel oil.

TABLE 2-11. EFFECTS OF CARBON MONOXIDE*

CO in the air — % by volume	Effects
0.02	Possible mild frontal headache after two to three hours.
0.04	Frontal headache and nausea after one to two hours; occipital (rear of head) headache after two and one-half hours.
0.08	Headache, dizziness, and nausea in forty-five minutes; collapse and possible unconsciousness in two hours.
0.16	Headache, dizziness, and nausea in twenty minutes; collapse, unconsciousness, and possible death in two hours.
0.32	Headache and dizziness in five to ten minutes; unconsciousness and danger of death in thirty minutes.
0.64	Headache and dizziness in one to two minutes; unconsciousness and danger of death in ten to fifteen minutes.
1.28	Immediate effect; unconsciousness and danger of death in one to three minutes.

*Muroka, J.S., Technical Report 144, U.S. Naval Civil Engr. Lab. 1961

TABLE 2-12. POLLUTANTS YOU GENERATE — LBS/YEAR

Fuel Need per year	Dry Hardwood 6 Cords	Anthracite Coal 4.5 T	No. 2 Fuel Oil 750 gal	Natural Gas 100,000 ft³
Particulate	420	54	3	1.5
Sulfur Oxide	—	85	31	.05
Carbon Monoxide	100	405	1	1
Carbon Dioxide	38,000	21,150	19,500	116
Hydrocarbons	65	11	1	.5
Nitrogen Oxides	140	20	30	12

3 BASICS OF DESIGN AND INSTALLATION

Before selecting a central heating system you should understand how the basic systems operate and how you determine what size unit you need. With several hundred manufacturers building these units, design, construction, and quality vary widely. The greater your understanding, the easier your choice will be.

Equipment Definitions

A *furnace* differs from a stove or heater in that a sheet metal jacket encloses the firebox to capture the heated air and direct it through ducts to the rooms in the house (fig. 3-1). Today most furnaces have a thermostatically controlled blower to move the air. Many older gravity systems depended on the principle that heated air becomes less dense and rises, while the cooler air falls to the floor and then through a return duct to the furnace.

In the better furnaces, the firing rate or the amount of heat produced is controlled by a thermostatically controlled damper or blower. Safety devices to dampen the fire should the furnace overheat or the power go off are also included.

A *boiler* is basically the same as a furnace except that the fire heats water either to a high temperature around 180° F. or to steam (fig. 3-2). In layman's terminology a boiler is often referred to as a furnace when discussing the home heating system. Heat is distributed to the house through a pipe and radiator system that contains the hot water or steam.

In the hot-water system the temperature of the water in the jacket surrounding the firebox is controlled by an aquastat that adjusts the draft on the fire. When the water gets too hot, the damper closes, and conversely when the water cools, the damper opens to allow the fire to pick up. The comfort level of the rooms is controlled by a thermostat that operates a circulating pump to move the heated water through the radiators.

In a *steam system,* when water turns to steam at 212° F. or higher, depending on the pressure, it absorbs a lot of heat called latent heat. The steam is then forced through the pipes and radiators by the pressure maintained in the boiler and controlled by a pressuretrol that operates the draft system. In

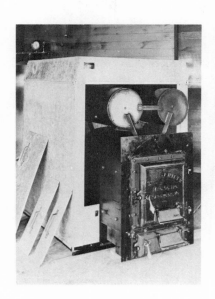

Fig. 3-1. This Daniels model is a typical hot-air furnace.

Fig. 3-2. Northland boiler model DF 620 is a typical boiler.

traveling through the radiators, the steam releases the heat and condenses back to water. The water flows or is pumped back to the boiler to be reheated.

Although steam boilers are still manufactured by several companies, they are not installed very often except as replacements in existing steam systems. Here are several disadvantages:

1. Steam responds less to changes in heat demand as it is either flowing or not flowing.

2. The installation is more critical than with water systems as pipes have to be pitched properly and traps and valves located properly so they don't cause water hammer, a noise caused when steam comes in contact with condensate water.

3. More attention has to be paid to the operation of the boiler to see that the correct water level is maintained and the controls are operating properly.

Fig. 3–3. Northland's Stelrad NC–2 is an add-on wood/coal boiler.

Fig. 3–4. This multifuel unit has an oil burner (lower right) plus solid fuel burner (at left).

The *add-on furnace or boiler* is a solid-fuel device that is attached to an existing central heating system to supplement that system (fig. 3-3). Many units can be installed as either independent units or add-ons. The advantage of this type of installation is that the cost is greatly reduced because the existing distribution system is used. This can provide full heating or just supplement your present system but in either case if the wood or coal fire goes out, the conventional system takes over automatically.

Some precautions have to be taken in making this type of an installation. It has to be sized properly so that it is compatible with the existing system. Because of the potential for overheating, pipes and ducts may require additional clearance or be insulated to meet the installation code requirements. Some manufacturers and some codes require that the add-on be connected to a separate Class A chimney flue. A second flue in the basement is seldom found in the more recently built homes.

If you are building a new home or replacing the heating system in your present home, you may want to install a *multifuel furnace or boiler* (fig. 3-4). There are units available that use oil, gas, or electricity and burn wood or coal as the solid fuel.

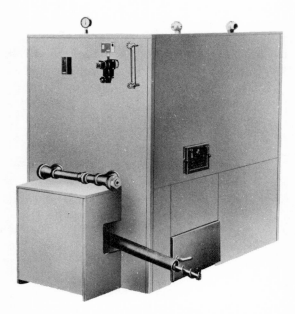

Fig. 3–5. A stoker for use in larger buildings, this Van Wert unit can be used with either steam or hot water systems.

Although multifuel units cost more to install, they have the advantages of simpler connections to the distribution system and automatic switching from one fuel to the other. In some units the solid fuel is ignited by the oil or gas burner. Most units have a single flue connection, making installation simpler.

One problem with some units is the fouling of the burner by the soot or creosote from the solid fuel fire. To overcome this, some manufacturers use separated or totally independent fireboxes.

A *stoker* is a device that feeds a solid fuel into a combustion chamber, provides the proper amount of air for burning the fuel under automatic control, and in some cases removes the ashes automatically (fig. 3-5). The stoker must be designed for the type of fuel to be burned; buckwheat or rice anthracite coal, stoker-size bituminous coal, wood chips, or sawdust.

Fuel can be burned more efficiently if fed by a mechanical stoker rather than by hand because the stoker provides a uniform rate of fuel feed, better distribution in the fuel bed, and positive control of the air supplied for combustion. It also reduces the work of tending the fire as the fuel hopper is large enough to hold at least a week's supply.

Although many new furnaces and boilers can be purchased with the stoker already installed, it can also be adapted to many existing units with some modification. Controls are similar to other heating units with the exception of a hold-fire timer that operates the stoker occasionally to keep the fire burning even though no heat is required.

TYPE OF HEAT

Heating systems can be classified as warm air, hot water, or steam. For reasons discussed in the last section, steam is not generally installed unless it is a retrofit to an existing system.

Your choice is simple if you are adding a solid-fuel central heater to an existing system. With about half of the total cost being in the pipes and radiators or ducts, it doesn't make sense to change the system that you already have.

On the other hand, if you are planning a new home or adding central heat to reduce the cost of keeping your electrically heated home warm, careful consideration should be given to what you install. Each type has its advantages.

Warm-air systems are the most common, because of lower cost and the adaptability to central air conditioning. Other advantages include:

1. Rapid response time when the furnace is turned on

2. Easy to add humidity

3. House is provided with filtered air

4. Fewer cold spots in rooms if adequate air returns are installed

5. Registers take up less room space than radiators

6. Doesn't need draining if the house is left unheated during the winter.

Hot-water systems have been gaining in popularity since the development of the baseboard radiator. Even though the initial installation is more expensive, the advantages may be worth the difference. These include:

1. Quieter operation — the pipes don't create the noise that ducts do

2. More uniform room temperature because of the residual heat of the water in the radiators

3. Less space is needed in the basement for the pipes than for ducts

4. The domestic hot water is easier to install in a boiler although it is available in some furnaces

5. Energy can be saved by lowering boiler water temperature in the fall and spring

6. No objectional room air movement as with some warm-air systems.

Within each classification there are many variations and options. Before making your selection, seek the advice of a competent heating contractor, stove shop, or engineer.

MATERIALS

When you start looking for a central heating unit you will find that they vary in the materials that were used: cast iron, steel, and firebrick. An understanding of these materials and why they were selected by the manufacturer will help you evaluate the quality of each unit.

Steel

Steel is used in the construction of most units. Its low price, combined with great strength, allows its use in the parts that receive the greatest stress. Steel can be formed into complex shapes and into large pieces to eliminate many of the joints found in areas such as a cast iron firebox. It can also be welded easily.

For material less than 3/16 inch thick, steel is designated by a gauge number. The lower the number, the thicker the sheet steel. For example, sixteen gauge = 0.06 inch or about 1/16 inch, eleven gauge = 0.12 inch or about 1/8 inch. This material is used in baffle plates, outer shells, stovepipe, and ducting. Plate steel is thicker material and often used in the form of 1/4-inch or 5/16-inch boiler plate for the firebox, water jacket, and heat exchanger. Steel tubing and pipe may also be used. These are designated by gauge or thickness.

Thinner steels are more apt to warp from the heat of the fire. This can affect the fitting of doors and can stress welds until they eventually break. The high temperatures developed in the firebox will cause oxidation of the steel, hastening its deterioration. That is why better quality units are made with a firebrick or ceramic liner.

Cast Iron

Many old-line furnaces and boilers and most units imported from Europe are made from cast iron. This material is iron combined with carbon and silicon. The parts such as the firebox, water jacket, doors, and dampers are cast from molten metal poured into sand molds. After cooling, the castings are cleaned and machined to fit tightly together. Asbestos rope, furnace cement, or high-temperature fiberglass is used to seal the joints between sections. A check of the castings for thin spots, large sand pits, tight joints, and warping will give an indication of the quality of the unit.

Cast iron is often used because it can take higher temperatures, such as from a coal fire, without oxidizing or burning out. Many cast iron furnaces and boilers have been in use for a hundred years or more.

One disadvantage of cast iron is that it is more brittle than steel. Sudden temperature changes such as opening the fuel door rapidly and allowing a lot of cold air to enter may stress the metal so that a crack could develop — and it is almost impossible to stop a crack from lengthening. Expansion and contraction from the heating and cooling will keep it working. Drilling a small hole at the end of the crack or having a good welder braze the joint may work for a while.

Most cast iron fireboxes should be taken apart every few years to replace the seals at the joints, otherwise air will leak in and you will reduce efficiency and the ability to control the fire. The stove bolts holding the sections together may have to be chiseled off and replaced. The joints should be scraped and sanded lightly to remove the old joint sealant and a new layer of furnace cement applied before reassembly. Furnace cement is available at most hardware stores and stove shops. It should be allowed to dry for at least a day before a fire is made.

Most manufacturers use cast iron for their grates. These are made so that they can be easily replaced if they burn out or crack. Replacements can be purchased from the manufacturer.

Cast iron and steel have about the same heat transfer characteristics. The rate of heat transfer depends on the thickness of the metal. That is why a thin sheet metal stove tends to give heat more quickly than the thicker cast iron one.

Refractory Materials

To take the high temperatures developed in the firebox area most manufacturers use firebrick, fireclay, or a ceramic material. A lined firebox has several advantages over an unlined one.

1. The refractory material protects the metal, extending its life.

2. It helps to retain a more even fire and therefore more uniform heat output. It also helps to force the heat to the heat exchanger.

3. Insulating type refractory materials allow the unit to be placed closer to combustible materials.

Refractory materials are made from clay containing silica, alumina, and iron. They can be manufactured to take temperatures up to 3000° F. These materials are available in brick, sheet, and cast form and can be made to fit all types of fireboxes.

When used as firebrick it is either cemented in place or held with metal brackets or retainers. Sometimes it is cast to the shape of the firebox.

With hard and constant use most refractory materials deteriorate over a period of years. The bricks in a furnace can be replaced by purchasing new ones available at most stove shops or heating supply stores. A cast liner is more difficult to replace because the furnace has to be disassembled. If the manufacturer has changed designs, the liner may no longer be available. Have a mason see if individual firebrick can be cut and cemented in place.

WHERE TO BUY YOUR FURNACE

In 1980 there were more than 200 manufacturers and importers of central heating units. A few have been in business since the 1800s but most have started since the energy crunch. Many manufacturers started out making stoves and, seeing that market become saturated, expanded in the whole house heating market. Because these units require more design to meet safety and environmental standards and more capital to finance the business, some manufacturers are already falling by the wayside. Keep this in mind as you shop for your new central furnace or boiler.

Today, central heating units can be purchased at many stove shops and alternate energy stores, or from a heating contractor. Most places will carry one or two lines, not the ten to twenty lines frequently found in the stove market. The reason for this is the larger investment needed and the lower demand. Furnaces are not found in department stores, hardware stores, or lumber yards, mainly because this type unit is not something you buy and install yourself, like a stove.

It's best to shop around until you find a dealer who carries a good selection of units, can supply accessories and parts, and provides a qualified installer. You can locate the dealers near you by referring to the telephone directory or weekly advertising newspaper, or contacting some of the manufacturers and distributors listed in the catalog section of this book. Before buying, inspect a unit for quality and the desirable features mentioned throughout this book. Also ask for the names of homeowners who have had the unit installed, and either call or visit them.

Most manufacturers work through regional distributors who stock the furnaces, parts, and other supplies that the local dealer may not have space to store. They also provide training programs and technical assistance. The European units are brought in by importers who serve the same function as distributors. In purchasing import units you will pay the duty fee and the added transportation cost. Most of the imports have been manufactured for many years and are of high quality.

WARRANTY

A manufacturer's confidence in the quality and safety of his product shows up in the warranty. It is only one year for some while others may replace or repair parts for up to fifteen years.

The unit must have been installed properly to have the warranty in effect. Most manufacturers require that it be done by a licensed contractor. Second, the furnace must have been operated using the suggested fuels and without overheating.

The warranty usually does not include transportation or labor costs associated with replacing the defective part. It excludes items such as firebrick, door gaskets, thermostats, blowers, and paint. The manufacturer's purchased parts such as the controls often carry their own warranty.

Your best bet is to purchase a quality product from a reputable dealer. His reputation is often more important than the warranty.

EFFICIENCY

In your visits to furnace dealers and your review of the manufacturers' literature, the high efficiency rating of a particular unit is often pointed out. Sometimes these ratings approach the 90 percent level. What does the efficiency mean and is it of any value to you as the purchaser? There are three types of efficiency usually associated with heating units. You should understand which of these is being discussed.

COMBUSTION EFFICIENCY — The percentage of heat that is released during burning a particular fuel. Most heating units can attain 70–90 percent efficiency.

HEAT TRANSFER EFFICIENCY — The percentage of the heat that is absorbed by the air or water in the heater. This is the biggest limitation in most units.

OVERALL ENERGY EFFICIENCY — The combination of the previous two. It is the ratio of the heat that is supplied to the home divided by the amount of heat that was available in the fuel. A low efficiency indicates that either the fuel wasn't burned completely and the heat went out the chimney as smoke or into the ash, or the heat was not absorbed by the water or air and was lost up the chimney.

Efficiency tests can be used as a guide to selecting between two units provided the same method and the same test standards were used. Don't expect to achieve the same high efficiency in your own home. Most tests are run using kindling-size pieces of kiln-dried wood. The average efficiency for full season heating will be in the range of 45 to 60 percent for wood-fired units and 50 to 65 percent for coal-fired units. A few units incorporating the new design concepts such as heat storage or gasification may exceed these values.

DRAFT REGULATION

Draft is the movement of the gases through the furnace caused by the imbalance created either by the heated gases becoming less dense and rising as in a chimney or by a difference in pressure created by a fan. For the fuel to burn, oxygen is supplied from the air entering below the fire (primary air) (fig. 3-6). The exhaust gases are carried away by the chimney. The greater the amount of air that is supplied, the higher the rate of combustion. Secondary air needed to burn the volatile gases and other combustibles given off from the fire should be supplied just above the fuel bed. It's most efficient when this air is introduced in high-velocity streams that give good mixing.

The draft needed to get good burning is affected by the following:

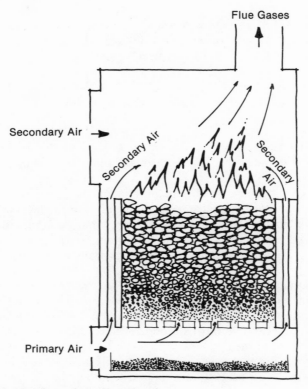

Fig. 3-6. The flow of air through a furnace

1. Kind and size of fuel. Coal requires more than wood; smaller pieces, because there is less space between them, require more than large pieces.

2. Thickness of the fuel bed. The thicker the bed the more resistance to the flow of the gases.

3. Resistance of the furnace-chimney system. Soot accumulations, multipass boilers, small fire tubes, and stovepipe connectors with several elbows require a stronger draft. Some manufacturers specify in their literature or installation instructions the minimum draft needed for their furnaces.

If your installation does not have adequate draft, poor burning may result, the grate may have to be shaken more often, and in some cases the fire will die out.

The three common ways of obtaining draft are with a natural draft chimney, a forced draft blower located below the grate, or a draft-inducing fan in the stovepipe or chimney (fig. 3-7). In the natural draft chimney, air is supplied through dampers in the ash door and feed door on most units. A hotter fire, taller chimney, or an insulated chimney will tend to increase the draft.

Very few manufacturers of furnaces and boilers use manual draft regulators such as adjustable slides, caps, or flaps that are common on stoves. These require more attention to get uniform heat output and high efficiency.

Thermostatically operated dampers are more common and can be of two types. In the first, a heat-sensing element is located on the unit in an area where it is not affected by air movement in the room. As the thermostat expands and contracts it controls the position of the damper. On some units both primary and secondary air is controlled by one thermostat. Top and bottom dampers are tied together by a rod and adjusted as the thermostat moves.

To get better sensing of the temperature in the living area, a conventional room thermostat can be used to activate a motorized damper on the furnace. Several damper settings can be achieved by using a multistage thermostat.

The water temperature in the jacket around the boiler is maintained using an aquastat that adjusts the damper setting. The aquastat is usually a bimetallic strip (two metals with different expansion rates) that moves a chain connected to the draft door.

A more recent method uses a small blower that forces air into the area below the grate. A thermostat or electric aquastat turns the blower on or

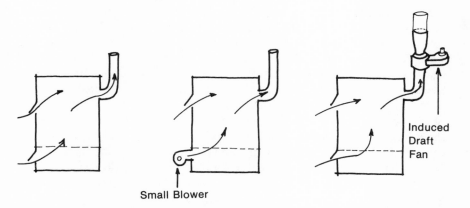

Small Blower

Induced Draft Fan

Fig. 3-7. Units are, from left, natural draft, forced draft, and induced draft.

off. A minimum air supply, either through a slide damper or other device, is needed to keep the fire going when the thermostat is inactive for long periods of time.

SIZING THE UNIT

When the home is warmer than outside, there is a continual loss of heat from the building, and the heating system must supply sufficient heat to maintain the desired temperature. This heat loss takes place in several ways. Usually the greatest loss is through the ceiling unless it is well insulated. Other large losses occur through the walls and windows, and by infiltration of air through cracks.

The calculation of heat loss from each room is important because it is used to size the radiators, piping, or ducts. Unless you have had experience with this, it is best left to the heating contractor or installer. These calculations can get very complex when you consider all the aspects of heat loss.

Here are two simple ways of calculating the size of furnace you need, to help you when you go shopping and also so you can talk intelligently to the heating contractor.

The *heat capacity of your present system* can be used as a guide for sizing a replacement or add-on unit. If your system is gas or oil, the heat output is usually listed on the burner or furnace label in Btu per hour. This could have been modified slightly by the furnace serviceman if you have built an addition to your home or if insulation has been installed since the furnace was put in. A check with the serviceman can often give you the answer.

In electrically heated homes, if you know the installed watts of heat or add up the wattage of all the heaters in the house and multiply the watts by 3.4, the Btu per hour capacity can be calculated. For example, if the total watts for the heaters in all the rooms is 13,000, the heat capacity is 13,000 watts × 3.4 Btu/watt = 44,200 Btu per hour. An additional 10 to 20 percent should be added if the furnace is to be placed in an unheated basement.

The *rule of thumb method* uses data based on an average home. The size of furnace needed is related to the floor area, an average heat loss factor for the home, and the section of the country where you live. Using the map in figure 3-8, locate the winter design temperature where your home is located. Next, refer to table 3-1 to get the heat loss factor base on the level of insulation present. Finally, calculate the total floor area of the home. Add in the second floor area if your home is two stories. Now multiply the floor area times the heat loss factor to get the approximate size furnace needed.

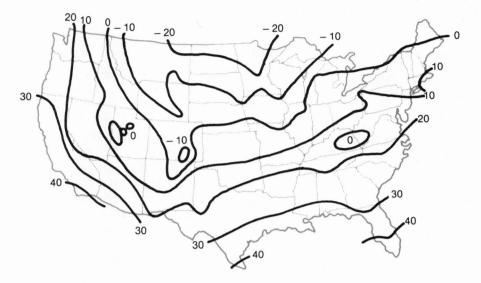

Fig. 3-8. Use this winter design temperature map to estimate heat loss.

For example, a poorly insulated home in Chicago has a heat loss factor of 104. If the downstairs floor area is 1,000 square feet and the second floor has 700 square feet, the heat loss on the coldest day will be:

$$1,700 \times 104 = 176,800 \text{ Btu per hour}$$

These methods should be used only to give an approximate furnace size because local climate, house design features, and other factors may have an overriding influence on the heat needed. Also most solid-fuel central heating systems can be installed 10 to 20 percent smaller than conventional systems because the heat output is continuous.

TABLE 3-1. HEAT LOSS FACTOR

Winter Design Temperature	Insulation Level		
	Poor	Medium	Good
−20° F.	136	85	34
−10	120	75	30
0	104	65	26
10	88	55	22
20	72	45	18
30	56	35	14

Getting Your Unit Installed

The installation of a furnace or boiler is quite different from a stove. The clearance and stove pipe installations are similar, but the electrical and plumbing work requires more skill than most homeowners possess. For this reason the majority of manufacturers emphasize that their units be put in place only by a qualified installer. In most states this means a heating contractor with a license. In some states this is required by law, and an installation will not be approved by a building inspector or fire marshal until a licensed heating contractor has certified it is installed to meet the code.

Most dealers selling furnaces and boilers have someone who can do the installation, either someone on their staff or a local contractor. This is probably your best choice as they will be familiar with the installation of these units and will often give you the lowest price. It may pay to get a quote from one or two other contractors just to see that the installation cost isn't way out of line.

One way to reduce the installation costs, which usually run from one-quarter to one-half of the total installation price, is to work a deal with the installer so that you can help with some of the less skilled jobs. These include cutting holes in the floor or walls for pipes and ducts, running the electrical wiring, and fitting the stovepipe.

If you are skilled in most aspects of the installation, you may want to do it yourself and then pay a contractor, if you can find one, to inspect your work and sign the certification form. You will have to familiarize yourself with the codes and obtain the necessary permits.

In hiring an installer you should inquire about the liability insurance he carries and what guarantee he gives on his work. Usually the guarantee is for six months or a year. A contract should be signed specifying the materials to be used and the work to be done. This protects both you and the installer. You are dealing with the safety of your family, and a good installation should be your primary concern.

Placement

There is usually one best location for your furnace. This can be decided with the help of the heating contractor. The best location in most homes is in the basement. The full benefit of the heat is then used as it travels up from floor to floor. With the extra height the chimney will provide a better draft that is needed, especially if you are burning coal. Some units can be installed on the same floor as the living space but special provisions for the ducts or water pipes will have to be made.

A central location is usually best for even distribution of the heat but you will have to consider other factors first.

1. If the unit is to be an add-on, it's best located near the existing furnace. This way, connecting pipes and ducts will be short.

2. A separate Class "A" chimney flue is required for most units. Do you have this available in your present chimney or can a separate chimney be conveniently located?

3. Is there space for the furnace, including at least four feet in front for firing and ash removal, three feet on one side for accessibility to dampers and cleanouts, and three feet above for pipes and ducts?

4. Will the unit fit through the basement door or bulkhead?

5. Can the fuel be reached conveniently?

6. Will there be at least a five-foot clearance to any fuel oil tank, paint, or other flammable material storage area?

7. Does the basement floor stay dry? If not, you may have to pour a concrete pad to set the furnace on.

8. Is the electrical supply adequate and convenient?

9. Can the furnace be placed where children can't come into contact with hot ducts or pipes?

10. Is the area free of explosive dust or volatiles as from a workshop?

Codes and Regulations

Concern for safety has kept some homeowners from installing a solid-fuel central heating system. In reality a central system is usually safer than a stove because of the added controls. If properly installed and operated, there is no greater danger from a solid-fuel furnace than there is from an oil or gas furnace.

Safety standards have been developed for the manufacturers of solid-fuel furnaces and boilers, but, because the industry is fairly new, they are not as well developed as those for other heating appliances. These standards are legally binding and will stand up in court. They may change from time to time as new methods or techniques are developed.

Although most of these standards are voluntary, state or local regulatory agencies can require that units meet these requirements. This is the case in the industry at this time. In some states a furnace or boiler must be listed by

ENERGY TESTING LABORATORY
of MAINE APPROVAL

A.I.R.I.
APPROVED

ATLANTIC INDUSTRIAL
RESEARCH INSTITUTE

Fig. 3-9. A code identification may be attached to your furnace.

a recognized testing laboratory before it can be sold or installed. In others it just has to be approved by the local inspector. What is the difference?

LISTED. This means that the furnace has passed the strict testing procedure of an accepted testing laboratory. The procedure consists of setting up the unit following the manufacturer's installation instruction manual in a room in which temperatures can be recorded in more than fifty locations on and around the furnace. The several types of burn and smoke tests include a flash fire test in which the drafts are left wide open. If the furnace passes these tests, it is listed in a product directory with any special requirements for installation. A check is made during and after manufacture to insure that quality and workmanship standards are maintained. You can be reasonably sure that you have a safe unit if it has been listed.

APPROVED. These furnaces or boilers are acceptable to a local or state building inspector or fire marshal for installation in your home. Tests may have been conducted on the unit or it may just have been inspected.

NONLISTED OR NONAPPROVED. Because of the high cost of getting a heating unit tested (several thousand dollars), many small manufacturers have not taken this important step. Most of their units are safe and would pass the test, possibly with slight modifications. Despite this, the inspector in your area may not let you make the installation. I know of several cases where units have had to be removed for this reason.

Several states with building codes require that a furnace or boiler be listed before it will be approved by the building inspector. Some require that it be installed by a licensed heating contractor. Check with your local authorities before you purchase a unit.

At times you may find a conflict between codes or standards and how they are interpreted. In general, if the installation manual has been approved by the listing agency, it supersedes other codes. Also the interpretation of regulations is generally left up to the local inspectors.

There are also standards for such things as fuels, chimney construction, and smoke emissions. During a recent coal shortage a neighbor of mine received a five-ton shipment of anthracite coal. After having much difficulty in getting the coal to burn, he had it analyzed in a laboratory and found the coal exceeded the acceptable ash content by several percent. He complained to consumer protection officials, and the dealer was forced to remove the coal.

Building Permits

In many states a building permit must be obtained before you can install or add to a central heating unit. Some states have a statewide code that is followed. In others, city or town codes cover the installation. All are based on nationally recognized codes or standards such as those of the National Fire Protection Association, the Council of American Building Officials, and the American Society of Mechanical Engineers.

Check with your local building official, town hall, or city manager's office to see whether a permit is needed. In areas that don't have building codes, the local fire marshal or fire chief often makes the inspections. In most cases the installer will obtain the permit for you.

Where a permit is required, a small fee, about $5 to $10, is charged for the inspection. This is probably the best insurance that you can buy. A signed inspection permit is valuable should you have a fire and need to collect from your insurance company.

The insurance company that issues your homeowner's policy may also require notification. These companies like to be told when a heating system is changed, especially when the move is to a solid fuel. Some companies just require you to check a box on your renewal form; others will want to have it inspected either by their own agent or the local building official.

Inquire about any rate increase. Some companies require a rider be attached to your policy with an additional premium paid. Most companies have had very few losses from the use of solid fuels and have not been too concerned.

Combustion Air

To burn properly, solid fuels require from 150 to 200 cubic feet of air per pound of fuel. For most homes this amounts to about one-tenth of the volume of the air in the home each hour. Except in new homes with very tight construction and plastic vapor barriers, infiltration greatly exceeds this so no special provisions for make-up air are needed.

There are, however, a few situations in which additional air should be provided. These include a tight home with a fireplace operating much of the time. Fireplaces and Franklin stoves use much more air than an airtight stove or furnace. Also in underground and envelope wall homes, infiltration is held to a minimum. Under certain conditions make-up air will help to reduce venting problems with an oil or gas furnace or water heater operating at the same time as the solid-fuel furnace.

The use of a separate duct to bring in outside air and direct it to the area of the furnace is recommended by most codes (fig. 3-10). One square inch of opening should be provided for each 2,000 Btu per hour output of the furnace. PVC or stovepipe, insulated to prevent moisture condensation and with one end screened and the other containing an automatic damper, makes a convenient installation.

If the furnace is located in a separate enclosed room, make-up air must be provided either through an air duct to the outside or through louvers in the wall or door.

The claim that the use of outside air can increase energy efficiency by not using the heated air to operate the furnace has not been proven. The total efficiency of the unit is usually less as the fire runs cooler and heat transfer is reduced. Greater amounts of creosote from a wood fire may also result.

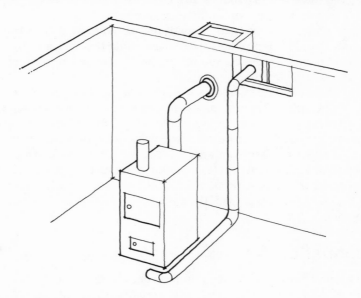

Fig. 3-10. One method of getting outside air into the furnace.

4 FURNACE SELECTION AND INSTALLATION

In the evolution of the movement back to solid fuels, the stove has generally made our homes less comfortable than when the oil or gas heating unit was used. The room in which the stove is located tends to overheat, and rooms farther away are often uncomfortable to live in. Use of several stoves has been tried but this requires separate chimneys and considerable work.

A stove heats the space around it by radiation and circulation, but adjacent rooms receive the heat only by the movement of the heated air through doorways. Furniture and other obstructions slow this movement so that the heat loss is greater than the heat gain. The furnace system overcomes this problem by supplying heat through ducts to each room.

Gravity Furnaces

In design, the furnace is basically a large stove that has been enclosed in a cabinet. The simplest ones are located in the basement and are called pipeless, having a large floor register placed directly above (fig. 4-1). This system was common in grandpa's home during the 1920–40 era. If you have ever seen one of these systems when it was "fired up," you may recall that you couldn't touch or even stand over the register for more than a few seconds because of the heat. Usually located under the center of the house, it did a good job of heating if circulation could take place. Upstairs rooms were heated by smaller registers in the floor. The cold air usually returned to the basement through the stairwell.

The method of controlling the fire was also very simple. The furnace damper was connected by chains to a wheel or lever placed in a convenient room upstairs. Temperature sensing was by body comfort; if you started feeling cold, you turned the lever that opened the damper.

This type heating system is still available from several manufacturers and will provide good comfort if your home is small, compact, and well insulated. Not having a blower, this unit will not add to your electric bill. One improvement that has been added is the use of a thermostat, damper motor, and limit switches to make operation more convenient and safer.

An improvement to the gravity furnace that allows its use in larger, more spread out houses is the addition of a plenum above the heat exchanger, and

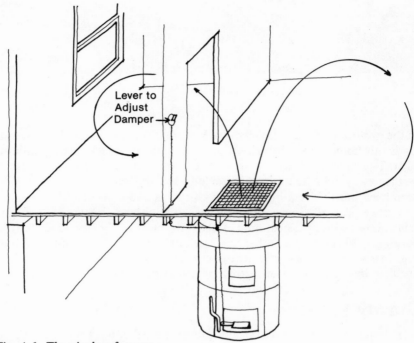

Lever to
Adjust
Damper

Fig. 4-1. The pipeless furnace

ducts to carry the heat to each room. If the furnace and the ducting are sized properly, rooms will be kept uniformly warm. Air is usually brought back to the furnace jacket through a centrally located floor register and duct. To operate efficiently, doors have to be left open. Even so floors may be 8° to 10° colder than the ceiling without much mixing.

Both the pipeless and the gravity furnace require a large air space between the firebox and the jacket to allow for air movement. This means the unit requires more space than a furnace having a fan. Air filters are not used because they would restrict air flow.

Forced Air Furnaces

Most furnaces have a blower to move the air past the heat exchanger and through the ducts (fig. 4-2). This increases efficiency about 10 percent over a similiar sized gravity furnace and gives more uniform temperatures within the living space.

Although a few manufacturers have taken their largest wood stove and enclosed it with a sheet metal cabinet, most are well-designed units based on

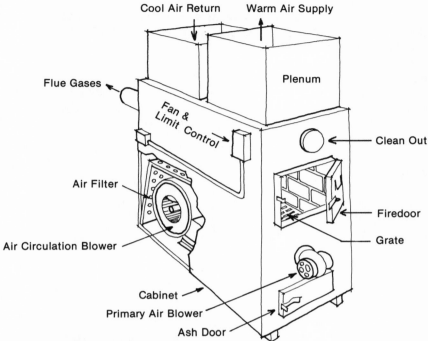

Fig. 4-2. The forced-air furnace

sound heat transfer engineering. Combustion takes place in the firebox where high temperatures can be maintained to keep the fire burning and capture the heat from the volatiles. The products of combustion then pass through the heat exchanger and are removed through the flue. Cool air from the living area is brought into the furnace through the return ducts, is heated as it passes around the heat exchanger, and then forced back through the ducts to the rooms.

Although most units are based on this principle, variations in design make some units cost more, weigh more, and operate more efficiently or safely. What should you look for when shopping for a furnace?

Cabinet

The cabinet encloses the firebox, heat exchanger, and usually the blower, and is made of light gauge cold rolled steel. It is usually painted with enamel of a color that distinguishes it from the products of other manufacturers.

Look for one to two inches of foil-faced insulation, either compressed fiberglass or mineral wool, attached to the inside of the cabinet. This will keep the heat within the furnace so that it can be carried to the upstairs living space. No need to heat the basement area unless you want to, and this can be taken care of with a separate duct or two with a damper to close when heat isn't needed. It will probably cost you an extra ton of coal or cord of wood a year to heat the basement. Insulation in the cabinet will also reduce the blower noise.

The cabinet should have access doors to allow you to change the filter, maintain the blower, and clean the fire tubes. For emergency operation without power to operate the blower, some furnaces require that doors be opened and filters removed.

Grate

The grate is one of the most important parts of the furnace designed to burn coal. Although wood can be burned without a grate, a coal furnace needs one for several important functions.

1. It supports the ash layer and the fuel above it.

2. It allows the removal of some of the ash layer without disturbing the fire.

3. It allows the primary air to be distributed evenly to the fuel layer and to pass up evenly around the pieces of burning coal.

4. It can be used to settle the coal down if it starts to cake or wedge itself into the firebox.

Almost all grates are made of cast iron. The quality of the casting and the care with which the furnace is operated will determine how long they will last. A grate can be warped or burned by shaking the fire so much that the hot burning coal rests on it. Most grates are designed so that they can be replaced. Some dealers and distributors stock replacement grates.

Grate design can affect the way a furnace operates (fig. 4-3). I receive many calls from stove owners having problems in keeping a fire burning for more than a day at a time. A poor grate design is often the cause.

A good grate will have the following features:

1. A system for shaking the whole grate area. Some grates have no shaker. This means that you have to insert a poker or "fiddle stick" to get rid of the ashes. This mixes the ash layer and the molten coal layer above it and forms clinkers. These clinkers then block the grate openings and restrict air flow.

Some grates have only a movable center section. Ashes tend to build in

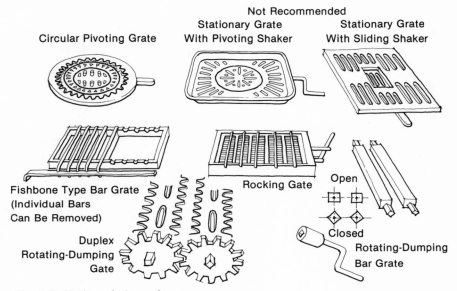

Fig. 4-3. Various designs of grates

the corners and restrict air flow. Often this stifles the fire and causes it to go out. It's a nuisance to have to remove the coal and start a fire every other day.

In a whole grate shaker system the grate can rotate around a center pivot, rotate in sections, or slide back and forth. An added feature with some grates is that they can dump the ashes and unburned coal into the ash pan. This is very convenient in the spring and fall when you may not want to operate the furnace continuously.

2. The spaces between the bars will be sized to support the size of coal that the furnace was designed for. Most manufacturers design the grate for the coal marketed in the area where the unit will be sold. A furnace may not work well if this coal is not burned.

3. The air space in the grate will be one-third to one-half of the grate area and should be evenly spaced. This supports the ashes and still allows good air movement. Some manufacturers make a different grate for burning wood as less bottom air is needed and the ashes are finer.

4. The bars are tapered so that ashes will not wedge between them.

5. A solid system supports the grate so that it can carry the weight of the fuel without stress.

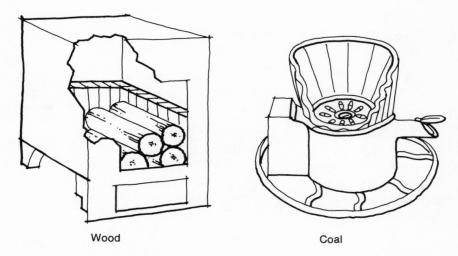

Wood   Coal

Fig. 4-4. Firebox designs for wood and coal

Ash Pit

The furnace should have a large ash area, preferably with an ash pan located below the grate. If you burn 100 pounds of coal a day, the average usage on a cold winter day, you will get almost two gallons of ashes. If the unit has a small ash pan, you will have to empty it each day. This leads to dust and ashes that have to be cleaned off the floor or stove pad.

The level of ash in the ash pit should be kept well below the bottom of the grate. This will insure that adequate air reaches all parts of the grate, and will prevent the grate from becoming overheated.

Combustion Chamber (Firebox)

It's easy to tell the difference between a furnace designed primarily to burn wood and one designed for coal (fig. 4-4). Just look at the firebox. If it is large and capable of taking a 24-inch chunk of wood, it was designed to be used mainly with wood. If it is round or small and compact, it was made for coal.

Generally there is no problem burning wood in the coal firebox except that it has to be cut into short pieces. In fact, wood should be used in the spring and fall when you want a quick fire to take the chill off the house.

You will have more difficulty burning coal in the firebox designed for wood even though the manufacturer may list it as burning both. The heat loss from the larger surface area may allow the temperature of the fire to drop below the ignition point, putting the fire out. This is especially true during mild weather. A full-open, full-closed damper system will not provide enough air to keep the fire burning. This means dumping the ash and coal and starting the fire over again.

The firebox should be designed to take the heat of the fire. A wood fire may reach 2,000° F., a coal fire, about 2500° F. Cast iron or special steel is used. Most manufacturers use a firebrick or refractory liner to protect the metal near the hottest part of the fire. This type of design is better than unprotected steel which will burn out in a few years from oxidation. Some units are listed for *wood only* because of the design of the firebox.

The firebrick evens the temperature of the fire, absorbing heat when the fire is hot and giving off heat to bring a new charge quickly up to temperature. It also helps to keep the fire going with the on–off automatic damper system commonly used.

A large firebox, one that will hold enough fuel for at least five to six hours in the coldest weather, is desirable. It should also have a feed door large enough to take ten- to twelve-inch chunks or the end of a coal scuttle.

Almost all furnaces are of airtight construction today to give better control of the draft. Check how this is accomplished and how difficult it is to replace the seal material. Asbestos, fiberglass, or other nonmelting material is generally used and is available from most dealers.

Look in the firebox area for a smoke curtain, a small metal hinged flap at the top of the fire door opening on some furnaces, to keep smoke from coming into the room when the door is open. Although it is not objectionable, it sometimes gets in the way when fueling, and is an indication that the furnace may have draft problems.

Heat Exchanger

The heat exchanger absorbs heat from the fire and conducts it to the air that is heating the house. Simple designs use a welded steel box that makes up the top part of the combustion chamber (fig. 4-5). Others use fins or baffles to get a larger surface area. Special shapes or designs with tubes or pipes extending through the exchanger increase the heat transfer area. A balance must be maintained to get maximum heat transfer without losing too much draft through the furnace.

A measure of the effectiveness of a design that is sometimes used is the ratio of heat exchanger area to grate area. Values of 15:1 to 25:1 are common with the lower ratio better for coal. This ratio is not commonly listed in manufacturers' literature but can sometimes be calculated from the information supplied. A ratio that is low would indicate low efficiency because the heat generated by the fire could not be absorbed quickly enough before

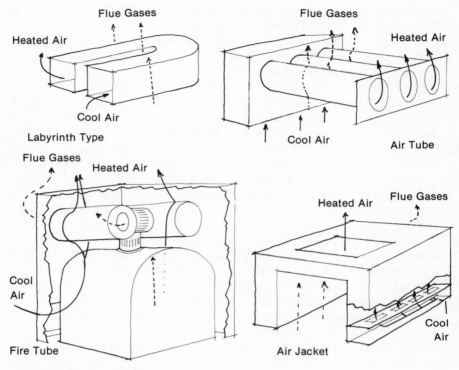

Fig. 4–5. Heat exchangers

the gases left the furnace. Too large a ratio would be efficient but could indicate possible draft or creosote problems.

Look for a design that has easy access to the heat exchanger for cleaning. Soot and fly ash that accumulate on heating surfaces act like insulation, reducing the heat transfer. Some manufacturers provide clean-out ports and special brushes for this purpose.

Size and Weight

Before purchasing a furnace, check to see that you can get it into your basement. You can imagine the frustration of getting the unit home and not being able to get it through the bulkhead. It has happened. Most units are designed to fit through the standard thirty-six-inch door. Larger units such as may be used in an old farmhouse may exceed this width.

Some manufacturers ship the furnace in several pieces, to be assembled on site. Parts like the bonnet, blower section, and oil burner, if a dual-fired unit, fit this category. The reason for this is to prevent damage during shipment and to reduce the weight.

If you had trouble moving your 400-pound stove into the house, consider how you will get the 500- to 1,000-pound furnace in. You will need plenty of help for this job plus some strong planks, rollers, and a hand winch. Plan this operation carefully before starting. On some units firebrick can be removed or is shipped separately, and doors can be removed to reduce the weight.

Blower

In a forced-air furnace system a blower is used to produce air flow. Of the two types available, the belt-driven blower is more desirable than the direct-drive type because it is quieter. It is also possible to adjust the output slightly by changing pulleys or motor should the need arise.

The key to an efficient furnace system is to have the blower sized properly. It must provide enough pressure to the air (called "head") to overcome the resistance of the filter, heat exchanger, and ductwork. It also must be capable of moving a large enough quantity of air (cubic feet per minute-cfm) to carry adequate heat to each room. Each model blower is tested and rated by its manufacturer for optimum performance.

Each furnace-duct system is different and the capacity and resistance should be calculated to determine the size of blower needed. If insufficient air flow or low pressure results, the heat generated by the fire may not be carried away fast enough. This could reduce efficiency, allowing more heat to escape up the chimney with the flue gases, or could activate the high limit control, shutting the damper. If the blower has excessive capacity, it may cycle often because it is removing the heat rapidly from the heat exchanger.

There is a wide variation in what manufacturers provide for a blower with their units. A survey of about fifty companies showed blower sizes from 265 cfm to 1,800 cfm for furnaces in the 80,000 to 100,000 Btu per hour class. Those with low output are not large enough to handle anything but a very simple duct system, maybe one or two registers. There can also be problems when the unit is used as an add-on and the heat sent through the existing duct system. More on this later.

There is also a significant difference in blower motor sizes. For example, consider two blowers each capable of 1,000 cfm output at the same pressure but one using a half-horsepower motor and the other only one-quarter horsepower. This translates to an operating cost of about fifty cents per day for the small motor and a dollar a day for the larger one. Why the difference? Mainly it is because of the fan design including the diameter and curvature of the blades. Look for a larger diameter blade.

A few manufacturers supply variable speed or multiple speed blowers. If connected to a multiple-stage thermostat, this could be advantageous.

Heat Distribution

Heated air is carried through a system of ducts to individual rooms. With proper design the temperature variation, air movement, and noise will be minimized. Because of the complexity of locating and sizing ducts, this should be left to a professional, either the heating engineer or contractor. Here is why and what needs to be considered.

1. Heat requirements of each room. This depends on size, exposed wall and window area, location of the room, and how well insulated the walls and ceilings are.

2. Duct size. Air speed and friction losses should meet accepted standards. Air speed may exceed 1,000 feet per minute (fpm) in main supply ducts but should not be greater than 400 fpm as it enters the room, otherwise you may feel a draft. Friction loss should be about the same in each branch.

3. Duct location. All duct runs should be as short as possible. Head space and the obstruction caused by their location in the basement should be considered.

4. Register and grille location. Selection of the right type of diffuser so that drafts are avoided is important in both the baseboard and high side wall register location. Location of return air grilles depends on whether the doors to rooms are normally open or closed.

5. Balancing. Proper location and adjustment of dampers is important to obtain uniform comfort levels. Heating contractors have special equipment for this job.

Duct System

Although several types of duct systems are common for distributing the warm air, the extended plenum system is most convenient for a basement installation (fig. 4-6). In this system a large rectangular duct extends out from the furnace plenum and generally in a straight line down the center of the basement ceiling. Round or rectangular supply ducts extend as branches from the supply duct to the outlets in the rooms. The large supply duct permits a better air flow with reduced resistance. The branch ducts can often be

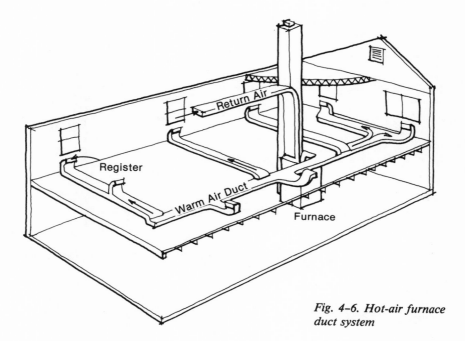

Fig. 4–6. Hot-air furnace duct system

located between the joists to take up less space. To reach second-floor rooms, riser ducts can be run through a closet or wall stud space. Use of the stud space is less desirable because there's room for little if any insulation.

Duct Material

It's important to consider the different duct materials available when selecting a system. Galvanized steel, most popular for many years, is being replaced with fiberglass. Fiberglass is quieter and holds heat better. Standards for duct materials are part of most codes. Hundreds of types and sizes of duct fittings to fit almost any situation are available from heating equipment suppliers. This has helped to reduce the time needed for installation. Return ducts can be made of almost any material including wood, except for the last two feet where they connect to the furnace, which must be fireproof.

Points to look for in a good installation include tight joints that don't rattle, large radius turns to minimize friction, and adequate support to keep the ducts from falling.

To cover the end of a duct where it enters the room one of the following is used (fig. 4-7).

Grille   Register   Diffuser

Fig. 4-7. These are used to cover end of ducts.

GRILLES. To deflect the air flow so that it doesn't blow on you. They can be installed in the floor or on the side wall.

REGISTERS. Similar in design and function to a grille but have an added damper to regulate the volume of air into the room. To shut off heat to a room, the damper is closed.

DIFFUSER. Used mainly in ceiling installations to deflect air flow over a large area.

Damper

Another device used to control the direction or volume of air flowing through the duct is a duct damper. Both manual and motorized dampers are available. *Manual dampers* may be used to vary the flow between winter heating and summer cooling if an air-conditioning unit has been installed. It generally requires a greater volume of air flow with the cooling system. *Motorized dampers* are often used in a system that has more than one zone. For example, one zone heats the bedroom area and a second the living area, each controlled by a separate thermostat. Proper type and location of the damper is important.

Duct Insulation

With more attention being paid to energy conservation, it is more important to insulate ducts in areas that do not need heat all winter. Examples include basements, crawl spaces, and closets. This is a job that the home-owner can easily do.

Fiberglass and mineral fiber materials are available for application to both round and rectangular ducts. Insulation should have an aluminum foil vapor barrier placed against the metal if cooling is part of the system. It can be held in place with adhesive, light gauge wire, or duct tape. Do not use the common fiberglass used for house insulation as it is not rated for the temperatures that may occur in a solid-fuel furnace. Also do not use paper-backed insulation. For most installations six to eight inches is cost-effective for colder areas of the country.

Clearance from Ducts to Combustibles

Consider the situation where, after removing the ashes from your furnace, you forget and fail to close the ash pit door. You now have a potential for a runaway fire. The high-limit control, which would activate when the temperature in the bonnet exceeds 250° F., would have no effect. The blower, which in some systems would start to run continuously, may or may not be able to cool the unit. For this reason and for occasions of power failure, the safe distance from the bonnet and ducts to wood and other combustibles is greater than for a conventional oil or gas furnace. It also means that if you are planning to use an existing duct system with an add-on unit, the ducts may have to be moved to allow a greater air space.

An alternate approach rather than moving the ducts is to use a clearance reducer (see table 5-2). This noncombustible material with the one-inch air space may save you some work in the area around the bonnet or first few feet of duct where the distance required is greatest. Also, if you insulate your ducts with the prescribed amount, although not specified in the code, this should give you adequate protection without the clearance reducer. Although occurrences of overheating are rare, I've seen charred wood around ducts. This makes me appreciate the need for these code requirements. For listed furnaces follow the installation manual distances. For all other units use the values in figure 4-8.

Balancing the System

After the system has been installed it should be tested to determine if the heat delivery to each room is adequate and the distribution uniform. You don't want to have one room 10° F. warmer than the next. Testing is done with an instantaneous reading thermometer and air speed meter. Duct and register dampers may have to be adjusted and vanes set to give the desired air flow. Care should also be taken to see that the registers don't become blocked with furniture or dust.

Humidifier

With a hot-air heating system, it is relatively easy to add moisture to the air. In older, not-so-tight homes it is difficult to keep comfortable levels of

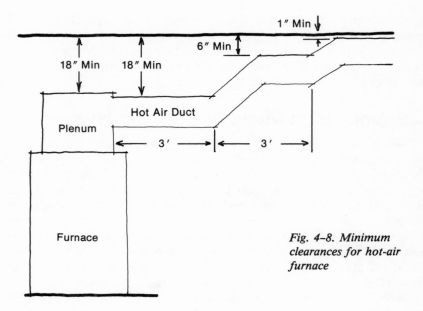

Fig. 4–8. Minimum clearances for hot-air furnace

humidity during the winter. Cold, dry outside air that infiltrates is often around 10 percent relative humidity level whereas the comfortable zone is 20 to 30 percent or higher. Some moisture is produced by cooking and bathing. This rapidly escapes, especially if a solid-fuel heater is used.

A pot of water is often placed on a stove to add humidity. With furnaces, a pan located in the bonnet is most common. Water level can be automatically maintained by a float. To increase the evaporation surface, porous plates may be used.

More sophisticated systems, such as a spray nozzle operated by an electric solenoid or a bypass which diverts some of the heated air through a wet, porous pad, are available. Both of these can be controlled by a humidistat in the living area.

A higher winter humidity makes you feel more comfortable at a lower air temperature. This saves fuel. It can also result in fewer colds and respiratory problems because nasal passages and throats do not dry as much. Check to see if a humidifier is available with the furnace you are purchasing.

Air Filters

Many manufacturers offer an air filter system as part of the furnace. This can be either part of the furnace enclosure or purchased as a separate unit.

As air circulates through the house, it tends to pick up dust. In a closed system with the return duct connected to the furnace, the air picks up room dust and carries it to the furnace and then back to the room. In a system without a return duct, where the return air moves through a basement stairwell or floor registers, dust from the basement may get mixed in.

The most common type filter is the spun glass replaceable type. These are readily available at department stores. A few manufacturers supply the aluminum washable type. Stay away from furnaces without provisions for air filtering.

Controls

Furnace systems use a variety of controls to provide comfort and safety. Because furnaces, unlike stoves, are located out of the living area, codes and standards generally require more devices to prevent or deal with overheating. Furnaces are also insulated, and overheating could cause

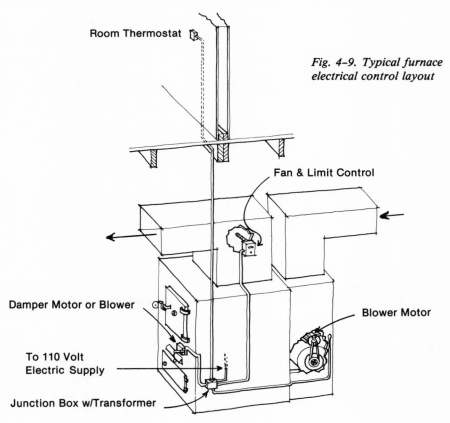

Room Thermostat

Fig. 4-9. Typical furnace electrical control layout

Fan & Limit Control

Damper Motor or Blower

Blower Motor

To 110 Volt Electric Supply

Junction Box w/Transformer

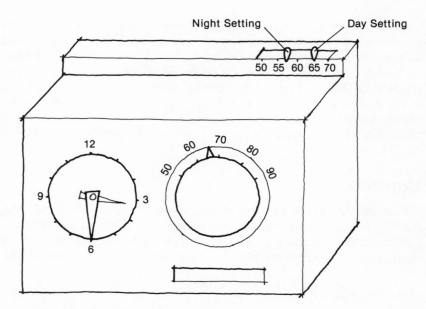

Fig. 4-10. Day-night thermostat

damage. Two basic functions needing to be controlled are the rate of burn and blower operation.

Control of the Fire

In most stoves the fire is controlled by slide or wheel draft controls. Few furnace manufacturers offer this manual method because it is difficult to keep an even heat output with it.

A second method uses the bimetallic thermostat to control the damper. This system is also poor because the firebox temperature is being sensed when what you really want to sense is the room temperature in the living area.

The third and most common system is similar to oil and gas furnace installations (fig. 4-9). A thermostat located in the living area, usually the living room, is used to regulate the fire. If the air temperature becomes too cool, the primary draft door is opened or draft blower activated and the fire picks up. Usually a damper motor or solenoid is used to open the draft door. Secondary air, if it isn't supplied through the primary draft, enters through a small manually controlled opening.

Locate the room thermostat in an area where it is not subjected to temperature extremes. An inside wall location about five feet above the floor and out of the path of duct or fireplace heat and sunlight is best. Low-

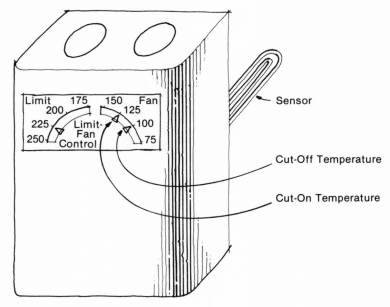

Fig. 4–11. Fan controls

voltage thermostats, supplied by some manufacturers, are recommended because they respond more quickly to temperature changes. Day-night or heat anticipator thermostats can be used for energy efficiency and comfort (fig. 4-10).

Control of the Blower

All furnace systems that have a blower to move the air through the ducts use a fan control. This device is used to turn the blower on and off in response to temperature changes in the furnace bonnet and prevents cold air from being circulated when the fire is low.

In the operation of the furnace, the room thermostat calls for heat and the draft is opened. The fire picks up and heated air rises through the heat exchanger to the furnace plenum. The fan controller is located in the plenum and is preset for a specific cut-on temperature (fig. 4-11). When the temperature of the rising air reaches the cut-on temperature, the blower is automatically activated and warm air is moved through the ducts. After a period of time the room thermostat is satisfied and shuts the furnace damper, slowing the fire. The air in the plenum begins to cool. When the temperature drops below the cut-off setting, the fan is automatically turned off.

Most manufacturers supply the fan control but it may have to be mounted when the furnace is installed. Cut-on and cut-off settings are adjustable and can be varied for seasonal comfort levels. Usual settings are between 130° and 150° F. for cut-on and 100°–120° F. for cut-off.

Just as in jet planes and space craft, a backup system is desirable. For the furnace, it is a fail-safe limit control, often combined with the fan control and called a fan and limit control. In case one of the controls fails and the temperature in the bonnet continues to increase above 200° F., the limit control takes over and shuts the damper and activates the blower. Also in event of power failure or shut off, the control will cause the fire to be checked.

Domestic Hot Water

An option on some units is a pipe or finned heat exchanger that is inserted between the jacket and furnace or in the plenum that will heat the water you use for cooking, bathing, and laundry. Installing this unit so that it will preheat water before it enters your present hot-water heater will reduce your energy costs and pay for itself in a few years. The heating of hot water is second only to space heating in home energy costs. A reduction in usage or a change in the type of fuel that you use can pay big dividends.

Accessories

What does the manufacturer supply with the furnace? It's best to check this thoroughly before you purchase one. Besides the furnace, you may need controls, cleaning tools, blowers and filters, heat reclaimers, and other devices. Some manufacturers provide everything; others, to make the price more attractive, recommend you obtain these locally. Add in everything when making price comparisons.

Converting a Converted Furnace Back to Wood or Coal

In the 1930s and 40s, when oil and gas were gaining in popularity, many solid-fuel furnaces and boilers were converted to those modern fuels. For oil, this conversion consisted of:

1. Removing the grate and the ash pit door

2. Reducing the size of the combustion chamber by lining it with refractory firebrick

3. Inserting the oil burner and then sealing with firebrick the front of the ash pit door through which the burner entered

4. Sometimes adding this firebrick in the area of the heat exchanger to increase turbulance and therefore heat transfer

5. Changing the control system

6. Adding a barometric damper to reduce the draft.

The gas-fired conversion burners usually came with their own refractory chamber so the firebox was not modified. These conversions often did not burn the fuel very efficiently. I have measured some that were as low as 55 percent.

There has been interest by some homeowners in converting back to coal or wood. What is involved?

First, see if you can find the grates, ash pit door, and controls. If you can, you're in luck. If not, determine who manufactured the furnace or boiler, not the burner. Look for a nameplate or a label on one of the castings. If you can locate the name of the manufacturer, you can find out if he is still in business. You might try checking with local heating contractors or the parts companies listed at the end of this chapter.

Second, if you have the parts or can locate them, see if you can find a heating contractor who will do the conversion. Someone who made the original conversions may be best. The burner will have to be removed, firebrick chipped out, and the unit inspected for damage. You may also want to have the furnace or boiler modernized to include a new control system and safety devices.

Third, if the manufacturer of the original unit can't be located, you may be able to adapt another grate system. You should expect to spend a minimum of $200 or more for the conversion. If on inspection you find that your unit is deteriorated and nearly burned out, it may be better to put the money toward a new system.

Checklist for a Safe Furnace Installation

Use this checklist as a guide for reviewing your installation before starting the first fire. It is intended to cover the principal safety requirements of the various codes. For a more detailed discussion of the installation procedures refer to the text.

1. The furnace is listed or approved for use in your state.

2. The installation/operation manual has been read.

3. The furnace is placed on a noncombustible floor, or adequate protection is provided for combustible floors.

4. List manufacturer's recommended clearances to combustibles.
   Front _____ Rear _____ Top _____ Side _____
   Cold Air Return _____
   Approved clearance reducers have been used where these distances have been reduced.

5. The following duct clearances have been maintained unless reduced by installation manual recommendations or clearance reducer.

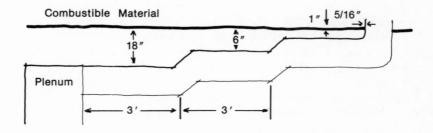

6. The chimney is clean and in good repair.

7. A UL-approved all-fuel factory-built chimney is used where a masonry chimney is not available or practical. It is installed according to the manufacturer's recommendations. (See Chapter 9)

8. Stovepipe twenty-four gauge or heavier is used. It is fastened at each joint with three sheet metal screws. (See Chapter 9)

9. The diameter of the stovepipe is not reduced between the unit and chimney flue. (See Chapter 9)

10. A draft regulator is located about eighteen inches from the furnace and adjusted not to exceed the draft in water column inches specified in the instruction manual. (See Chapter 9)

11. The length of the stovepipe is less than ten feet. No more than two elbows are used. (See Chapter 9)

12. There are at least eighteen inches between the stovepipe and combustible material unless a clearance reducer is used.

13. The stovepipe slopes up toward the chimney and enters the chimney higher than the furnace exhaust collar.

14. There is a tight connection between the stovepipe and the collar.

15. All heat ducts are rigidly connected and pitch upward away from the furnace.

16. The cold air return duct, if required, has been connected to the furnace.

17. The air filter has been installed.

18. The room thermostat is located where it is not affected by drafts, radiant heat, or duct air.

19. All electrical parts have been installed in accordance with applicable local and national codes.

20. The proper size fuses are used.

21. The recommended blower control settings, ON _____° F., OFF _____° F. from the installation manual have been checked.

22. The high limit switch is set at 250° F. or lower.

23. A metal container with a tight lid is available for ash disposal.

24. The building official or fire marshal has approved the installation.

25. The company insuring the building has been notified of the installation.

26. A smoke detector is installed near the ceiling in the area of the basement stairs.

27. A fire extinguisher (at least 2½-pound ABC type) is near the basement stairs or door.

Furnace and Stove Parts Companies

Aetna Stove Company, Inc.
SE Corner, 2nd & Arch Streets
Philadelphia, PA 19106

Avondale Stove & Foundry Co.
2820 6th Ave. S.
Birmingham, AL 35233

Brewer Atlantic Clarion Stove Co.
Brewer, ME 04412

Michael Brucher, Inc.
501 Homan Ave. N.
Chicago, IL 60624

Charleston Foundry Co.
161 Pennsylvania Ave.
Charleston, WV 25307

Empire Furnace & Stove Repair
793 Broadway
Albany, NY 12207

Portland Stove Foundry
57 Kennebec Street
Portland, ME 04104

5 BOILERS: SELECTION AND INSTALLATION

A boiler is a cast iron or steel pressure vessel that is designed to burn a fuel and to transfer the heat to water, the transfer medium (fig. 5-1). The *fire side* contains the fire box and is exposed to the products of combustion. The water in the *water side* absorbs the heat.

In a hot-water heating system the boiler, pipes, and radiators are always full of water. Air that develops in the system from the heating process must be vented. Provisions for drainage must also be provided. Although most systems installed today are closed systems, water occasionally has to be added. This is done automatically. Also, to take care of the expansion of the water when it is heated, an expansion tank partially filled with air is used.

As compared to the ducts in a hot-air system, the pipes used in the water system are small and do not take up much space in the basement. The water is moved through the pipes and radiators with a circulating pump. The radiators are usually placed along the outside walls, especially under the windows, to counteract the heat loss. Heat to the various sections or rooms of the house is regulated by the use of zoning, control valves on each radiator, or dampers on radiator covers. The controls required by code are designed to give a safe operation.

Design

Boilers are made either of cast iron or steel (fig. 5–2). With over 100 companies making solid fuel boilers, many designs and construction techniques have developed. A look at the advantages, disadvantages, and principles of operation of some of these may help you with your selection.

Cast Iron

These are made in sections which are assembled and held together by bolts or rods. Gaskets, cement, or a machined surface is used to make it watertight. The number of sections determines the heat capacity of the boiler. Larger boilers are assembled on site. This design has a long service life. If repair is needed, the defective section is replaced. Cast iron boilers

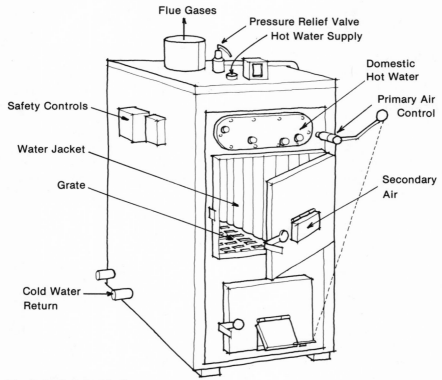

Fig. 5-1. A typical boiler

tend to be heavier than steel ones because thicker material is used. They are also more expensive.

Steel

These are welded together to give a specific size and heat rating. They can be an assembly of either horizontal or vertical tubes. In the *fire tube type* the flue gases pass within the tubes that are surrounded by water. They may be designed so that the gases make several passes before exiting through the flue connector. This concept is used to increase the efficiency of the boiler. However, the more pases the flue gases make, the greater the resistance, and either a strong natural draft or a draft inducer is needed to overcome this.

In the *water tube* type, the water passes inside the steel tubes while the hot flue gases pass over them. Smaller but more tubes are used than with the fire tube type. Several manufacturers use a down-draft principle with this design, so the flue gases are exhausted at or below the grate level.

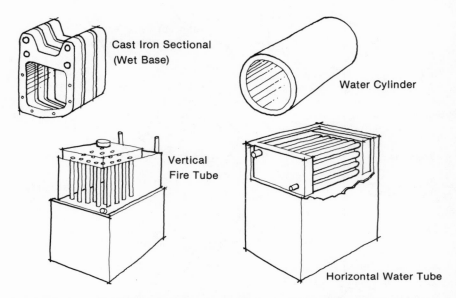

Fig. 5-2. Various designs of boilers, showing both steel and cast iron models.

Another style of water tube utilizes pipe formed into a cylinder that fits above the fire box. Water in the pipe is heated by the radiant energy of the fire and the gases flowing over it. This design, although simple and less expensive, has a smaller water capacity than most water tube boilers. Also, mechanical damage to the pipe can be greater as wood being fed to the fire can bump it. Life of the pipe before replacement has to be made is five to ten years, depending on use. This type is usually designed to operate as a combination boiler and space heater as it is rarely jacketed.

Many manufacturers use a *wet back* design either with or without horizontal fire tubes. A large water capacity is usually built in. This allows quick pickup of heat in the house as all radiators are filled with hot water. One disadvantage if the water jacket is allowed to enclose the firebox area (wet leg) is that it quickly absorbs the fire's heat, and poor efficiency and creosote problems can result. The inside surface of the jacket rarely exceeds 200° F. Most manufacturers recommend that only dry hardwood or coal be used.

A variation of the wet back design is called the wet base. By adding water-filled tubes as a grate or a water chamber below the fire, less protection is needed if the boiler is set on a combustible floor. Also a greater water capacity results. This type operates best if the unit is not oversized for the house and a hot fire is maintained.

One design popular in Europe and brought into the United States by

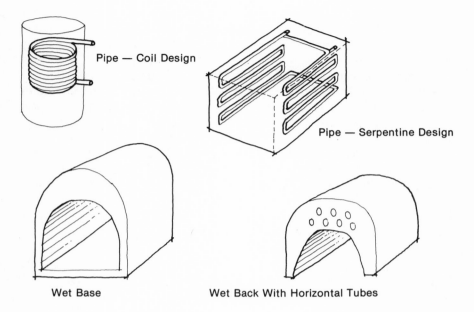

Pipe — Coil Design

Pipe — Serpentine Design

Wet Base   Wet Back With Horizontal Tubes

several importers is the *magazine feed* with vertical fire tubes. The fuel is placed into a vertical chamber and self-feeds the fire. Primary burning takes place at the base of the fuel supply, and secondary burning of the gases is in a separate chamber where high combustion temperatures can be maintained. The heat and gases then travel through the tubes. Some designs allow a reduction in the number of passes when starting or fueling the fire by using a baffle.

Evaluation of Design

A review of the manufacturer's literature will often give an indication of quality. If many specifications and dimensions are given, which is common with old-line companies and imports, you can be assured that these have resulted from tests and competition over the years. Some newer companies tend to give very few details and then often use the word "approximate." Sometimes the unit's performance is overstated.

Two measures of quality design can be applied if you can obtain the needed facts. If they are not listed, ask the dealer to get them for you.

1. Will the boiler provide the heat output that the manufacturer claims? Based on research by the U.S. Bureau of Mines, practical rates of heat transfer at design load for conventional boilers will average about 3,300 Btu

per hour per square foot of water jacket heating surface. The rate may be as much as 1,500 Btu per hour more at maximum load. To calculate the design boiler output for a boiler, multiply the square feet of heating surface by 3,300. For example:

25 square feet of heating surface × 3,300 = 82,500 Btu per hour

Boilers vary somewhat from this, based on factors such as the amount of surface that is direct surface (the fire shines on), which has a higher heat transfer, and the shape of the heat transfer passages.

2. Can the boiler burn the fuel fast enough to give the rated output? Combustion rates vary significantly with the type of fuel you are burning. Practical combustion rates for small boilers operating on natural draft are:

	A Efficiency of Combustion percentage	B Pounds of fuel per square foot of grate per hour	C Average Heat Output of Fuel Btu/pound
Anthracite, Pea	65	5	12,500
Anthracite, Nut	65	8	12,500
Bituminous	60	9.5	13,000
Wood	50	20	6,500

To calculate the amount of heat output based on the number of pounds of fuel that the boiler will burn multiply

GRATE AREA (sq ft) × A × B × C

For example:

1.5 sq ft grate × 65% × 8 lbs nut anthracite/sq ft-hr ×

$$12,500 \ \frac{Btu}{lb} = 97,500 \ Btu/hr$$

A boiler's grate area designed for wood may be too large when operated with coal. Other factors that can have an effect are the type of draft system and the size and shape of the combustion chamber.

Firebox

As with furnaces, the firebox should be separate from the water jacket so that the high temperature needed for complete combustion can be maintained.

Another advantage to this design is that the fire will recover quickly when the thermostat calls for heat. With the hot area maintained in the fuel, air entering through the draft will increase the rate of burn and achieve a hot, efficient fire in a couple of minutes.

The firebrick or cast iron liner is used by most manufacturers as a heat sink and tends to stabilize temperatures in that area. When a new charge of wood is put in, the fuel is dried and brought up to temperature quickly because it absorbs some of this heat.

Water Jacket Capacity

You may have noticed that oil- and gas-fired boilers have become smaller and smaller during the past few years. Higher efficiency burners and smaller water jackets are the reason. When the thermostat calls for heat, the burner turns on and almost instantaneously a large amount of heat is being transferred to the water. This is then pumped through the radiators and your rooms warm up.

With a wood or coal boiler, this process is much slower. In fact, by the time a conventional boiler has started, brought your home up to a comfortable level, and then shut off, the solid-fuel boiler may not be up to full fire yet. A large reservoir is advantageous then to have a supply of hot water ready when the thermostat calls for heat.

It is also needed as a heat sink because of the way that the solid-fuel boiler operates. Unlike the conventional burner in which the heat ceases when the fuel is cut off, the solid-fuel fire continues to burn and give off heat after the draft is reduced. A larger volume of water will absorb this heat without overshooting the aquastat set-point too much.

For boilers in the range of 80,000 to 100,000 Btu per hour, water jacket capacities vary widely, depending on the manufacturer and the design. The pipe-style heat exchanger units have capacities of one to ten gallons and should generally be used as an add-on to an existing boiler with a large water jacket or with an auxiliary storage tank. Most conventional water jacket units have capacities in the eight- to fifty-gallon range.

From fifteen to twenty-five gallons are required to fill the pipes and radiators of an average-size house (1,200 to 1,500 square feet). If old-style cast iron radiators are used, this may be more than doubled. Having a reserve capacity of hot water in the jacket will provide more even heating throughout the house.

Heat Storage Units

As discussed in the chapter on combustion, any boiler is most efficient when high firebox temperatures and good mixing of the air and combustion gases occur. Efficiencies of over 80 percent are possible. In the conventional method of operation this seldom occurs because the house warms up and the fire is dampered down before the high temperatures needed are reached.

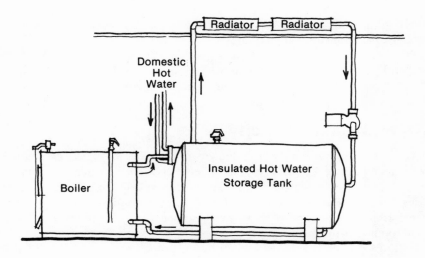

Use of a large water storage tank to hold heat for later use is an old idea used in industry and solar heating but now is applied to solid fuel heating (fig. 5-3). This system, although more expensive to set up, can be adapted to almost any boiler.

The system consists of a well-insulated tank, 250- to 2,500-gallon capacity, that is placed in the basement, garage, or outdoors. It can be of steel, concrete, or even wood and is usually not pressurized because of the cost. The water from the tank is circulated through the boiler and heated, then returned to the tank. After two to twelve hours of full-blast firing, the water in the tank will reach 150° F. or higher. Now you can let the fire go out because you have enough heat for from two days to two weeks, depending on the size of the tank and the time of year. A 2,500-gallon tank will hold about 1,250,000 Btu of usable heat. Heat from the tank is extracted with a heat exchanger and the hot water circulated through the radiators. The heat exchanger, which is usually a finned pipe, is needed whenever a non-pressurized tank is used. Advantages of the system include:

1. The use of fuel is more efficient, with very little creosote formed when wood is used.

2. You can leave home for several days and still have the house stay warm because the boiler does not have to be fired.

3. Although more expensive initially, the payback is shortened by the savings from increased efficiency.

4. It can be tied in easily with a solar collector system.

Fig. 5–3. Furnace and, on opposite page, the design developed by Dr. Richard C. Hill. Madawaska Energy Corp. is one of several firms manufacturing this type furnace.

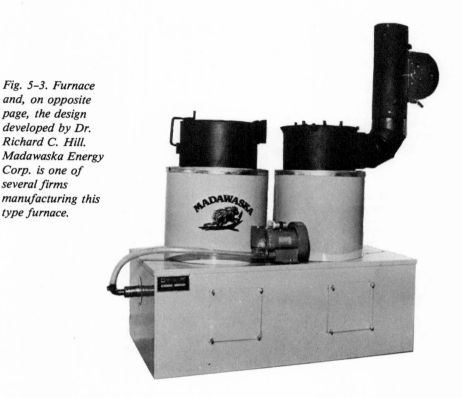

Since the original high efficiency furnace and the storage concept were developed by Professor Richard C. Hill at the University of Maine, several manufacturers have started marketing models using the concept. Homeowners who have installed the system have been pleased with the results.

Clearance for Combustibles

As with any other heating unit, the clearance to combustible materials must be adequate to give protection from overheating. With listed units the dimensions are specified in the installation instructions and these should be followed. For unlisted units use the local code requirements or the National Fire Protection Association recommendations shown in table 5-1.

There is a difference between a boiler that is jacketed by a tank of water around the firebox-heat exchanger area and one that has a set of pipes or a tank over only part of the area. The latter may lose much more heat to the air space around the boiler; therefore the safe distance is greater.

TABLE 5-1. BOILER SYSTEM CLEARANCES[1]

	Front	Sides and Rear	Top of Casing	Piping
Boiler all water wall or jacketed	48[2]	6[3]	6	1[4]
Boiler — not fully water walled or jacketed — example pipe heat exchanger	48	36	36	1

1. Adapted from NFPA recommendations.
2. This clearance considered necessary for fueling and maintenance.
3. Larger clearances may be necessary for access to piping and for servicing.
4. Clearance can be reduced to ½ inch where pipes pass through finished flooring, wall paneling, or ceiling. The gap should be covered with a noncombustible material such as a metal ring.

Spacers and Heat Shields

When necessary, clearances may be reduced if an approved material and method are used (table 5-2). All clearances should be measured from the outer surface of the boiler to the combustible material, disregarding any intervening protection applied to the combustible material. The one-inch air space is the key to the installation.

The spacer material and location are also important to prevent heat transfer. Noncombustible material such as porcelain electric fence insulators, pieces of steel conduit, thin wall pipe, or several small washers can be used. The heat shield should be fastened to the wall with screws or lag bolts located no more than twenty-four inches apart. This will keep the shield from warping and reducing the one-inch clearance. Spacers should not be placed directly behind the boiler. The heat shield could also be built free-standing with legs or suspended from the ceiling by a chain or wire.

If the boiler or furnace is to be located in an area that is finished and used for living space, such as the basement rec room, you may want to consider an alternative clearance reducer to make the installation more attractive. Although not listed by NFPA the following are generally acceptable.

1. Substitute copper or aluminum sheet for the sheet metal. Check with a local scrap metal dealer or use strips of roof flashing riveted together.

2. Cover one of the approved materials with a facing of tile, noncombustible artificial brick, or a fire-resistant paint. Use only a high

TABLE 5-2. CLEARANCE REDUCERS FOR FURNACES & BOILERS

| | Where the required clearance with no protection is: | | | | | | |
| | 36 inches | | 18 inches | | | 6 inches | |
Type of Protection	Above	Sides & Rear	Above	Sides & Rear	Stove pipe	Above	Sides & Rear
28 gauge sheet steel spaced out 1 inch	18	12	9	6	9	2	2
3½-inch-thick masonry wall spaced out 1 inch	na	12	na	6	na	na	2
22 gauge sheet steel on 1-inch mineral wool batts reinforced with wire mesh or equivalent and spaced out 1 inch	18	12	4	3	4	2	2

*Protection shall be applied to and cover all combustible surfaces within the distance specified as required clearance with no protection. Spacers and ties shall be of noncombustible material. Adequate ventilation shall be provided behind the protection. Mineral wool batts shall have a minimum density of 8 lbs/cu ft and minimum melting point of 1,500° F. If a stovepipe passes through the masonry wall there must be at least ½ inch of space between the pipe and the masonry. There shall be at least 1 inch between the appliance and the protector. Adapted from NFPA 211.

temperature silicone adhesive such as one labeled No. 103 or 106. This is available in a cartridge and is applied with a caulking gun.

3. Use commercial stove mats. Generally more than one are required to get the necessary size. Be sure that they are UL listed.

Floor Protection

The boiler or furnace is usually placed in the basement where the floor is either concrete or gravel. As both of these are noncombustible, no floor protecting is needed. However, if there is any possibility of flooding or if the floor is damp, the unit should be raised off the floor and set on bricks or four-inch concrete blocks.

If the unit is to be placed on a combustible floor, two other concerns should be recognized. First, will the floor support the weight? Most wood floors are designed to support about forty pounds per square foot. A furnace or boiler may weigh between 400 and 1,200 pounds. This requires special bracing and a weight distribution pad under the unit. It's best to consult a contractor or building official if you have this situation.

Second, the combustible floor must be protected from overheating and from sparks from the fire. A basic protector that is safe for almost all units is a layer of four-inch hollow masonry blocks laid with the ends unsealed and the joints matched in such a way as to provide free circulation of air through the masonry (fig. 5-4). The layer of bricks should be placed on or covered by a piece of twenty-four gauge sheet metal.

Protection on the side with the door should extend out at least eighteen inches. This protection can be the same construction as is under the stove or can be ¼-inch asbestos covered by twenty-four gauge sheet metal. If the unit is not enclosed with sheet metal or a water jacket, the floor protection should extend out the sides and back at least twelve inches.

Controls

A description of the basic boiler controls and their function will help in understanding how the system operates (fig. 5-5). These controls should be installed to handle overfire conditions and power failures as well as normal operation. They should also be installed to meet the National Electric Code and local codes.

THERMOSTAT. In a hot-water heating system the room thermostat starts and stops the circulating pump that moves heated water through the radiators. This is different than in a furnace where the thermostat controls the draft on the fire.

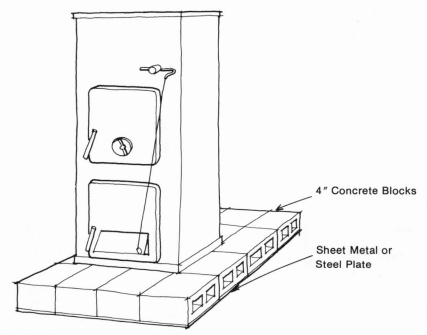

4″ Concrete Blocks

Sheet Metal or
Steel Plate

Fig. 5-4. Protector for a combustible floor

The house may be divided into zones, each with its own thermostat. This saves energy and allows different comfort levels. Each zone has to have its own circulating pump or control valve.

The room thermostat is usually located on an interior wall about five feet above the floor. Avoid areas that may be affected by heat from radiators, fireplaces, or the sun shining through a window.

AQUASTAT. This is a thermostat that is inserted into or attached to the boiler water jacket. Because of the difference in water temperature between the top and bottom of the jacket (10-25° F.) it is usually located near the top. Two types are used.

The *mechanical* type, usually containing a bimetallic sensing device, is connected by a chain or cable to adjust the draft door. When the water in the jacket cools, the damper opens to increase the intensity of the fire. After the water is brought back up to temperature, usually 180° to 190° F., the damper is almost closed. The chain connected to the damper should be adjusted so that if the water temperature reaches 190° to 200° F. the door is almost completely closed. A mechanical aquastat will continue to function even if the power is off.

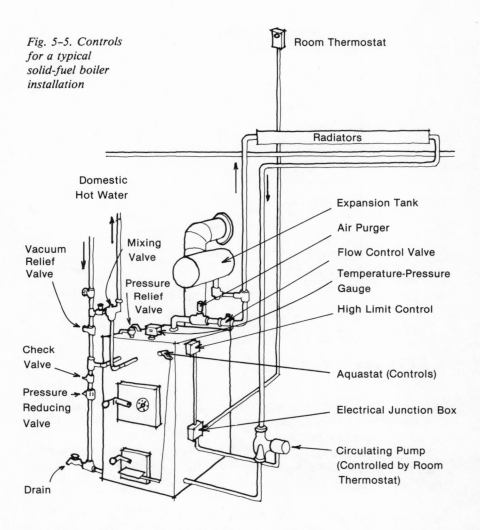

Fig. 5–5. Controls for a typical solid-fuel boiler installation

Room Thermostat

Radiators

Domestic Hot Water

Expansion Tank

Air Purger

Flow Control Valve

Temperature-Pressure Gauge

High Limit Control

Vacuum Relief Valve

Mixing Valve

Pressure Relief Valve

Check Valve

Aquastat (Controls)

Pressure Reducing Valve

Electrical Junction Box

Circulating Pump (Controlled by Room Thermostat)

Drain

The *electrical* type may use a bimetallic or thermister switch to operate a damper motor, solenoid, or blower. With this method the draft is either full open or full closed. A setting of 180° to 190° F. water temperature should be used. A second switch within the aquastat will only allow the circulating pump to start if the temperature of the water is above a set point, usually 150 to 170° F. This keeps the pump from running even if the thermostat calls for heat, should the fire be near the end of the burn cycle or be out. Otherwise cool water would circulate through the radiators. Both temperatures should be adjusted downward in the spring and fall when less heat is needed. A savings in fuel will result as there is less heat loss from the boiler and less overrun of the room set point.

Some aquastats have a high limit switch that activates the circulating pump and closes the damper should the high limit set point be reached. This is usually between 190° and 210° F., but the manufacturer's recommendations should be followed.

TEMPERATURE-PRESSURE GAUGE. This gauge is mounted on the top or front of the boiler for a visual check of water jacket conditions. Normal temperature should be below 180° F. at less than twenty pounds per square inch pressure.

PRESSURE-TEMPERATURE RELIEF VALVE. Should the pressure in the piping system or water heater exceed thirty pounds per square inch or the temperature exceed 225° F., this valve will open and release some water. Piping from the valve should be placed so that it is within four inches of the floor or connected to a drain. A pressure relief valve is usually located on the boiler.

PRESSURE REDUCING VALVE. Located on the cold water supply line to the boiler, this valve automatically adds water when the pressure in the system drops below a certain level, usually twelve pounds per square inch.

FUSIBLE SAFETY PLUG. Found as part of the safety system on a few boilers, it melts and releases a small amount of water or steam into the firebox to dampen the fire. Some manufacturers have stopped using this device because of leakage problems.

BLEEDER VALVE. Sometimes located at the top of the boiler or piping system to vent the air out of hot-water systems. As water is heated, trapped air is released and collects at high points in the system, causing poor flow and noise problems.

FLOW CONTROL VALVE. Prevents water from thermosyphoning through the system when the circulating pump is off. It also retains domestic hot water in the boiler during the summer when heat is not needed.

PREWIRED CONTROL PANEL. A few companies supply prewired harnesses or control panels to make installation faster and simpler. These may contain relays, transformers, temperature indicators, and other devices. Some use electronic rather than mechanical components.

LOW WATER CUT-OFF. This float switch, which is required on most steam systems and is optional on water systems, is connected to the damper to shut the fire down if the water level in the jacket falls to a dangerous level. It protects against the possibility of steam buildup and damage to the water jacket and pipes. These need to be flushed monthly and cleaned yearly to remove dirt and scale deposits.

STEAM GAUGE. Located on a steam boiler, it indicates the pressure of the system. Most boilers operate on low pressure of less than five pounds per square inch.

STEAM POP VALVE. A safety valve that allows the system to blow off steam should the pressure exceed fifteen pounds per square inch.

Expansion Tank

Water heated from 40° F. to 200° F. expands by about 4 percent of its original volume. In a hot-water heating system, provision must be made for this expansion, otherwise the pipes will burst or the boiler explode. The expansion tank with its air cushion serves this purpose (fig. 5-6).

Open expansion tanks are not used much today because they have to be located above the highest radiator, and this usually means the attic where freezing temperatures can occur.

The closed expansion tank, usually located near the boiler, must be sized to take the expansion of all the water in the system. This includes the water in the boiler, pipes, and radiators.

Sizing the tank properly is complicated and should be calculated by the heating contractor. Generally it is sized for at least 10 percent of the system water capacity for single-story homes and 13 percent for two-story homes. A typical ranch house having an 80,000-Btu-per-hour boiler with a fifty-gallon jacket might use a fifteen-gallon tank.

Heat Distribution System

The heat developed in the boiler can be moved to the room radiators by either natural circulation (gravity) or by pump. *Natural circulation* is practical only if you have a small, well-insulated house. Although you don't have the expense of the pump and the electricity to run it, pipe sizes and radiators must be larger because of the slower water flow rate. Also, unless designed properly, rooms will not heat uniformly.

Most systems installed today are *forced circulation*. They contain circulating pumps, piping, radiators, and control valves. Zone control, where

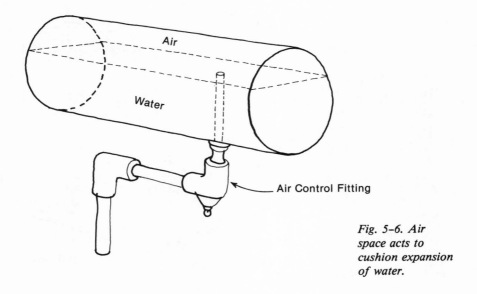

Fig. 5-6. Air space acts to cushion expansion of water.

different sections of the house have their own thermostat, can be used to save energy. Let's look at each of the system components.

Circulating Pump

This small centrifugal pump is usually located on the cool water return line next to the boiler. It provides the force to move the hot water through the pipe and radiators. In sizing the pump, the maximum amount of heat that is needed by the house and the temperature difference between the supply water and return water must be considered. For example, a ten-gallon-per-minute pump could be used on a 100,000-Btu-per-hour boiler with a 20° F. temperature differential. The rate of water flow in the pipes is usually kept below four feet per minute to keep friction losses and noise levels down.

Piping

Copper assembled with soldered joints is the most common material used. Although expensive as compared to steel, it resists corrosion and is available in a soft form that can be bent to fit into tight corners. It is available in three wall thicknesses: Type M, thin wall for solder fittings only; Type L, medium thickness, rigid or flexible, for most heating system applications, and Type K, heavy wall, rigid or flexible, for underground service.

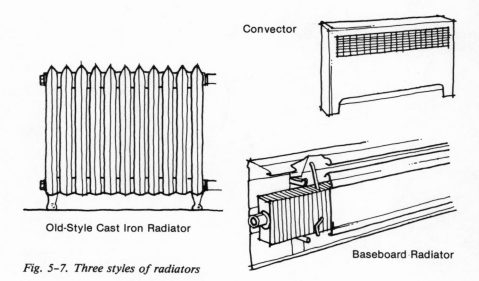

Convector

Old-Style Cast Iron Radiator

Baseboard Radiator

Fig. 5–7. Three styles of radiators

Supply and return pipe tappings on the boiler are generally between 1¼ to 2 inches, depending on the heat capacity. The main pipes supplying the radiators are generally no more than one size smaller. Larger pipe sizes provide adequate flow with low friction losses.

Piping must be installed to allow for expansion due to heating. One hundred feet of copper pipe will lengthen about 1¾ inches between a cold start and a 180° F. operating temperature. Straps or other type hangers should be used to handle this movement.

Heating pipes in areas that are not heated should be insulated. This includes crawl spaces, garages, and basements. If heat is needed in these areas, a radiator system with circulating pump can be installed. Fiberglass or polyurethane sleeves available at most hardware and plumbing supply stores are convenient to use. A less expensive material is the one-inch by six-inch by 100-feet sill sealer used under the foundation sill of modern homes. This can be wrapped or taped around the pipes. A double thickness (two inches) providing an R-7 insulating rating can be justified at today's fuel prices.

Radiators

Many styles and sizes of radiators and convectors are available (fig. 5-7). The bulky cast iron radiators common a few years ago have been generally replaced by baseboard radiators that occupy less space and give more uniform heat. Baseboard radiators are hollow cast iron or finned pipe units that replace the conventional baseboard. Located along the exterior walls of the house, especially under the windows, they counteract the heat loss that occurs there. They are available in standard lengths and in several heat outputs depending on pipe size and number and size of fins per foot. Lower output radiators are more desirable to give more uniform heating and less dust streaking on the wall.

Convectors usually consist of finned pipes enclosed in a cabinet with openings on the bottom and top. Hot water is circulated through the pipes. Cool air from the floor enters the bottom, is heated, and leaves the top, setting up circulation patterns within the room. Some units are available with fans to increase air movement. Convectors are located under windows or along the outside wall.

The number of radiators or convectors that are installed should be adequate to heat the room on the coldest day. This can be calculated by a heating engineer or contractor knowing the amount of window, wall, and ceiling area, how well the house is insulated, and the design temperature for your town.

Valves

Balancing valves are often used to adjust the flow of water to each room. An unbalanced system may allow one room to overheat while the next room is too cool. These valves can be located on the risers leading to each room or at other points in the stystem.

A *by-pass valve* may be installed to allow gravity operation during a power failure (fig. 5-8). This valve, which opens when the electricity goes off, is located in a loop around the circulating pump. A system designed for pumped flow can provide limited heat by gravity and, if the boiler is operated at a lower firing rate, can keep the house reasonably warm.

Piping Systems

The object of installing a central boiler system is to provide comfortable temperatures throughout the house. There are two general systems of piping that are used either with gravity or forced hot water. With either of these systems the distributing mains are located along the basement or crawl space ceiling with up-feed through the floor to the radiators. The forced system can also be used to supply radiators along the basement floor. Although there are many variations of the two basic systems, a simple explanation of each will be given here.

ONE-PIPE SYSTEM. One main pipe makes a circuit from the boiler around

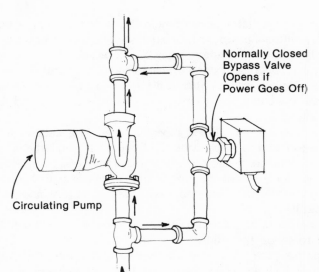

Normally Closed
Bypass Valve
(Opens if
Power Goes Off)

Circulating Pump

Fig. 5–8. Bypass valve is in loop around the circulating pump.

the perimeter of the basement and back again (fig. 5-9). This pipe serves both as the supply and the return. Two risers extend from the main to each room radiator. One riser carries the heated water to the radiator, the other returns the cooled water.

The one-pipe system has the advantage that it is less expensive to install than the two-pipe system. However, water cooled from each radiator mixes with the hot water flowing through the main, and each succeeding radiator receives cooler water. Although this may only be a few degrees, allowance must be made for this in sizing the radiators. Larger radiators may be required the farther along the system you go. This system works best in smaller homes.

Two-pipe system. This system has two main pipes. One carries the heated water to the radiators and the other returns the cooled water to the boiler. Although the supply water cools slightly from the first to the last radiator, heat loss is much less than with the one-pipe system. The *reverse return* layout, where the length of water circuit for one radiator is the same as for any other, is preferred to the *direct return* layout where short-circuiting can occur with the radiators nearest the boiler tending to overheat.

Zone Heat

In large rambling or multistory houses it is often desirable to heat one area less than another. For example, the bedroom wing, which is seldom

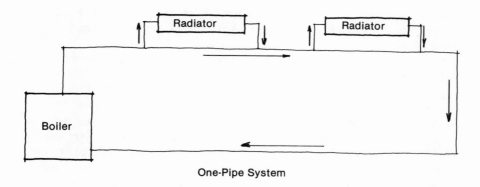

One-Pipe System

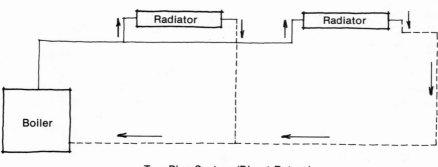

Two-Pipe System (Direct Return)

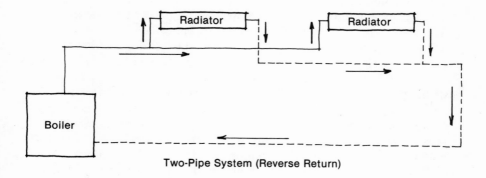

Two-Pipe System (Reverse Return)

Fig. 5–9. Three systems for circulating hot water

used during the day, could be heated to 55° F. rather than the 65° F. needed in the living area. This will save fuel, about 15 percent for each 4° F. temperature.

Several methods for doing this are used. The simplest uses flow control valves either on each radiator or on the supply line to each room. By reducing the flow, less water and therefore less heat is available for that room. The disadvantage to this system is that it is inconvenient to adjust, the valves being located in the basement. Its use is limited to seasonal setting.

A second system uses solenoid-operated *zone valves* activated by a thermostat located in each area. Each area must be piped as a loop but is supplied by a main circulating pump that runs whenever heat is needed in any zone. Although there is less maintenance because there is only one pump, more electricity is used than if each zone had a separate pump..

In the *multiple pump* system, each zone is independent of the other except that all zones are supplied by the same boiler. When the thermostat in a zone calls for heat, that zone pump is activated and runs until the temperature is brought up to the set point (fig. 5-10). This is the most desirable installation but costs the most.

In each of the last two systems, one zone has to be connected to the high-limit switch on the boiler to act as a dump should overheating occur. This is usually the largest zone. The smaller the water jacket capacity, the greater the chance for overheating. If a dump zone isn't provided, the pressure-temperature relief valve may trip often. This is undesirable.

Steam Heating Systems

Although steam heating systems are seldom installed today, there are still many in use in older homes. A brief description of how they differ from a water system may explain why water is more desirable.

Several manufacturers make solid-fuel boilers that can be operated with steam. Different controls including a pressuretrol, safety pipe valve, steam gauge, water gauge set, and low water cutoff are usually supplied.

The steam system operates with water at about the three-quarter level in the boiler. The water is heated to the boiling point, and steam is created. Pressure builds in the boiler and is controlled by the pressuretrol, which adjusts the primary air damper. When heat is needed in the living space, the thermostat opens a steam valve, and steam under pressure moves through the pipes and radiators heating them. When the steam cools off, it condenses to water and flows back through the pipes to the boiler where it is heated again. There are many variations of steam systems, each with different controls, piping, and valves. Both one-pipe and two-pipe systems are used.

Because steam is either flowing or not flowing, it responds less to changes

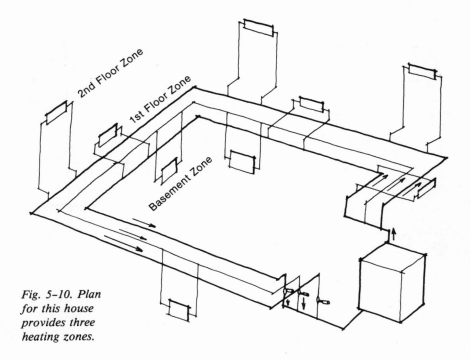

Fig. 5-10. Plan for this house provides three heating zones.

in heat demands. At the return end of each radiator is an automatic steam trap to keep live steam in the radiator and to allow the water from condensation to drip back into the return pipes. A thermostatic type steam trap is usually used in house heating systems. A condensate pump may be needed to return the water to the boiler. Installing or adding to a steam system requires specialized experience that most heating contractors have not acquired. If you have a steam system, you may have to search for someone who can do the job.

Domestic Hot Water

An accessory available for most boilers is a domestic hot water heating coil. With conventional system heating, it takes from 1 to 1½ gallons of oil or twelve to eighteen kilowatt hours of electricity to heat the hot water needed daily by an average family. With the solid-fuel boiler operating during the cold weather, it makes sense to utilize the less expensive solid fuel to do this job.

Two types of systems are commonly used (fig. 5-11). The *internal tankless heater* is immersed into the jacket water of the boiler. It is a

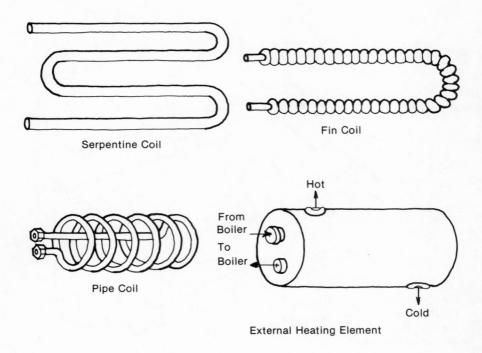

Fig. 5-11. External and internal domestic hot water systems

separate unit that bolts into an opening of the furnace jacket. Cold water flows through the coil and is heated by the hot jacket water that surrounds it. If the jacket water cools, the aquastat adjusts the draft and the fire picks up, heating the water again.Coils of various sizes are available but one that will heat four to six gallons per minute at 180° F. jacket water temperature should be adequate for the average home.

For boilers that don't have provisions for an internal coil, an *external tankless heater* can be used. A small (about five-gallon) well-insulated tank with a finned tube is mounted above the boiler. Hot water from the boiler jacket is piped into the tank so that thermosyphon circulation takes place. The domestic water circulates through the finned tube. External tankless heaters are available in many sizes and are a better choice if you need a larger hot water supply.

Both systems should be installed with a mixing valve so that water supplied to the house will be 120° to 140° F. The safety of not having scalding water and the energy conservation because of reduced heat loss make this a wise choice.

CHECKLIST FOR A SAFE BOILER INSTALLATION

Use this checklist as a guide for reviewing your installation before starting the first fire. It is intended to cover the principal safety requirements of the various codes. For a more detailed discussion of the installation procedures, refer to the text.

1. The boiler is listed or approved for use in your state.

2. The installation/operation manual has been read.

3. The boiler is on a noncombustible floor, or adequate protection is provided for combustible floors.

4. List the manufacturer's recommended clearances to combustibles
   Front _____ Rear _____ Sides _____ Top _____
   Approved clearance reducers have been used where these distances have been reduced.

5. A one-inch clearance to combustibles for all heating pipe has been maintained except one-half inch where it passes through finish flooring, wall paneling, or ceilings.

6. The chimney is clean and in good repair.

7. A UL-approved all-fuel factory-built chimney is used where a masonry chimney is not available or practical. It is installed according to the manufacturer's recommendations.

8. Stovepipe of twenty-four gauge or heavier is used. It is fastened at each joint with three sheet metal screws.

9. The diameter of the stovepipe is not reduced between the unit and chimney flue.

10. A draft regulator is located about eighteen inches from the furnace and adjusted not to exceed the draft in water column inches specified in the instruction manual. (See Chapter 9.)

11. The length of the stovepipe is less than ten feet. No more than two elbows are used.

12. There are at least eighteen inches between the stovepipe and combustible material unless a clearance reducer is used.

13. The stovepipe slopes upward toward the chimney and enters the chimney higher than the level of the boiler exhaust collar.

14. There is a tight connection between the stovepipe and the collar.

15. A pressure-reducing valve has been installed for boiler feed water.

16. All heat pipes are rigidly supported and installed to allow for expansion.

17. The room thermostat is located where it is not affected by drafts, radiant heat, or hot air.

18. All electrical parts have been installed in accordance with applicable local and national codes.

19. The proper size fuses are used.

20. Fail-safe controls are used.

21. The recommended aquastat temperature settings have been checked.
    High limit _____   Water temperature _____
    Circulating pump ON _____   OFF_____

22. A pressure/temperature gauge is installed on the boiler.

23. Where zoned heat is used, at least one zone is connected to the high limit for a dump.

24. The expansion tank is properly sized and connected.

25. A thirty psi pressure relief valve is installed on the boiler.

26. Pressure-temperature relief valves are installed on the piping system.

27. A metal container with a tight lid is available for ash disposal.

28. The building official or fire marshal has approved the installation.

29. The company insuring the building has been notified of the installation.

30. A smoke detector is installed near the ceiling in the area of the basement stairs.

31. A fire extinguisher (at least 2½-pound ABC type) is near the basement stairs or door.

6 THE ADD-ON

Most solid-fuel central heating units can be installed as add-on units, and some will work better when installed this way. The add-on system uses the existing distribution pipes or ducts of the oil or gas unit and allows wood or coal to be used as fuel rather than oil or gas (fig. 6-1). If you are already heating with hot air or water, the add-on furnace is usually your best choice. Here is why.

1. The installation cost is considerably less than if a whole new system is installed.

2. Controls can be arranged so that the conventional unit will switch on if the solid-fuel fire dies down or goes out. The temperature in the home is kept a comfortable level.

3. The conventional unit will provide the heat on cool, damp days in the spring and fall when only a little heat is needed and it is a bother to start a wood or coal fire.

4. A smaller solid-fuel unit can be purchased if it is to be used mainly as a supplement to an existing system with the conventional unit available to provide heat for peak heating needs.

As with all heating systems there are disadvantages too:

1. The add-on installation is usually more complex than a straight furnace or boiler. This results from additional controls, valves, and dampers that are needed to get the two units to operate together safely. Also many manufacturers do not give very clear installation instructions.

2. The space required for the two units is greater than for a single dual-fuel unit. If space is a limiting factor, the dual-fuel unit should be used.

3. Add-on systems do not work as well during power failures because of more resistance in the system.

Fig. 6–1. This add-on unit is Dover's Model W. B–30.

4. A separate chimney flue, often not present, may be required by the manufacturer's installation requirement or by code.

Here are the features to look for before purchasing:

SIZE. Generally the output of an add-on should not exceed that of the present oil or gas unit. Because the unit operates continuously rather than intermittently, less capacity is needed. Also, the distribution system, pipes or ducts, may not be able to handle more heat than comes from the present furnace.

As a guide, if the add-on has a well-insulated cabinet you should install a unit that is 15 to 25 percent less than the capacity of your present furnace. The furnace output is usually listed on the nameplate in Btu/hour.

If the add-on does not have a fully insulated cabinet, water jacket, or heat exchanger and part of the heat is transferred to the furnace room or basement, a capacity equal to the existing oil or gas furnace can be used.

CONTROL. Select a unit that can be installed to give good control of the fire and the heat output. Manual draft control requires more attention than does thermostatic control. The higher cost of a thermostatic system will be repaid by the savings in fuel in a short time. Safety controls including high-limit switches and safety valves should be included.

INSTRUCTION MANUAL. Often you can judge the quality of an add-on by the amount of detail in the instruction manual. A good manual should give detailed instructions on several types of installations that will meet the code. This is best done with drawings showing the components and how they are connected to the conventional system.

The manual should also go into detail on how to operate the unit at different times of the year, using different fuels and when the power fails. Because each model is slightly different and most homeowners have not had prior experience with solid fuels, these instructions are very important.

Before you purchase, it is best to buy or borrow the instruction manuals from the two or three units that you are most interested in and study them in detail. You are making a major investment in time and money and you want to have a unit with which you will feel comfortable and safe.

EMERGENCY OPERATION. You should know whether the add-on can be operated when the electricity is off. One of the reasons many homeowners install a solid fuel unit is to have heat during winter storms when the power may be interrupted.

Usually under this condition the unit operates by gravity or thermo-syphon action when heated air or water rises in the supply pipes and cooled air or water flows back through the return pipes. To operate this way efficiently, oversize pipes or ducts are needed and should be considered before the installation is made. Sometimes other steps such as the removal of filters, opening of dampers or valves, and the reduction in the firing rate have to be taken when the power fails. The procedure should be worked out, written down, and placed where it can be found quickly when the lights go off.

Furnaces

Details for installing an add-on furnace vary widely and depend on the location and set-up of the conventional system, the design and size of the add-on unit, and the local codes. For this reason it is strongly recommended that you hire a licensed heating contractor familiar with these units to do the installation. Most manufacturers also stress this point in their instruction manual and most void a warranty if the unit fails from improper installation.

A review of some of the installation procedures is given here to give you an understanding of the ways in which an add-on can be attached to your present system so that you can intelligently discuss the options with your heating contractor. It is not intended to detail how a unit is installed.

SERIES INSTALLATION. The easiest although the least desirable installation is adding the solid-fuel furnace downstream from the conventional unit (fig. 6-2). The cold air takes the normal path through the oil/gas furnace and then, through modification of the ducting, is sent through the add-on before entering the supply ducts. Only one blower is used. It is on the conventional furnace, and the original one has to be replaced with a larger model or modified by increasing the motor size and blower speed to overcome the added resistance of the air traveling through the new furnace.

Other points to consider include:

1. The add-on should be large enough to take the output of the conventional unit without creating too much back pressure; otherwise the blower motor may burn out or the unit will trip off from overheating.

2. Ducting between the two units and to the supply duct should be as large or larger than the present system. Also, large radius curves should be used to reduce turbulence and friction in the air flow.

3. A second thermostat has to be installed, preferably next to the present thermostat so that the same room temperature is being sensed. When wood or coal is being burned, this system thermostat should be set at the desired room temperature and the conventional furnace thermostat set about 10° F. lower. This allows a time lag of a few minutes to give the fire a chance to pick up once the damper is opened. If the thermostat settings are set too close together, the conventional unit will start up about the time the solid-fuel unit starts to heat up. This wastes fuel and may overheat the ducts because the output of both furnaces is combined in a single duct system.

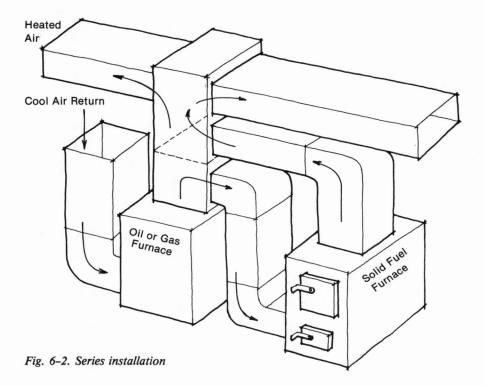

Fig. 6–2. Series installation

The difference between the two thermostat settings should be changed for different times of the year with the greatest difference used in spring and fall when heat loss from the house is relatively slow. Another way to handle the problem of both units operating simultaneously is to install a switch that will disconnect power to the conventional burner when you know you will be home to keep the solid-fuel furnace operating.

4. The fan control and limit control on the solid-fuel furnace have to be connected to the conventional furnace blower so that when the temperature reaches the set point in the bonnet of either furnace, the blower will start.

There are several reasons for not installing the solid-fuel furnace upstream from the oil/gas unit, although it is recommended by several manufacturers.

1. The blower motor would receive the heated air from the solid-fuel furnace and could overheat and burn out, or the thermal overload could be tripped.

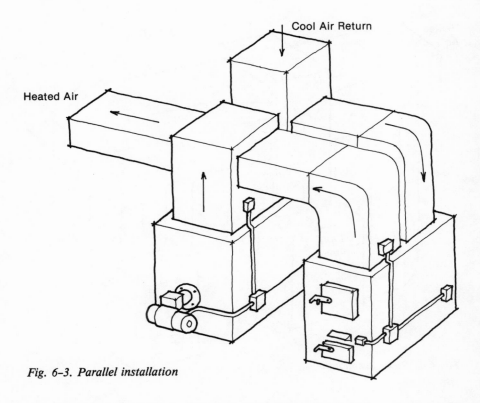

Fig. 6–3. Parallel installation

2. The blower in the conventional furnace is creating a negative pressure (suction) on the wood/coal furnace, and if cracks or holes develop in the heat exchanger, toxic fumes could be pulled into ducts and distributed in the house. This, although unlikely when the unit is new, is a possibility after a few years when the metal starts to burn through or cracks develop in the welds.

3. There is a greater chance for reverse flow during power outages when the blower can't operate as there may be less resistance to air flow back through the return ducts. With no clearance restrictions on return ducts, the combustibles could become overheated.

PARALLEL INSTALLATION. In this type of installation, the two units work independently of each other and their only connection is at the supply duct and the return duct. There is less heat loss to the inactive unit and less chance of damaging blower motors and controls. This is the preferred installation.

There are several ways of making the duct connections, depending on the layout of the present system. One of these is shown in figure 6-3. Y connections rather than T connections should be used if possible to reduce the turbulence and to prevent heat reversal through the non-functioning unit.

If the units are closely matched in heat output and blower capacity, the solid-fuel unit can be installed to operate independently of the oil/gas unit. The two room thermostats are located together with one wired to control the draft on the wood/coal furnace and the other set at a lower temperature, connected to start the oil/gas burner. The fan-limit control in each is wired to start the blower in that unit. No air moves through the oil/gas unit when the wood/coal unit is providing adequate heat.

If the add-on is smaller or its blower output is less than that of the conventional furnace, they should be connected so that the blower on the wood/coal furnace is removing the heat from the heat exchanger and moving it to the supply duct from the conventional furnace and the larger blower is distributing the heat throughout the house. A relay is needed to allow both blowers to operate at the same time. An increase in air velocity may be noticed from the combined output of both blowers.

The parallel installation usually works better than the series type during power failures because resistance to air flow is less. The manufacturer's recommendations should be followed but usually the filter and side panels have to be removed.

Semi-parallel installation. This variation is used with solid-fuel add-on furnaces that don't have provisions to connect to the cold air return duct (fig. 6-4). Air for this style furnace is taken from the basement or furnace room. Although one less duct is needed, these disadvantages should be noted.

1. Cold air has to return to the basement through an open stairwell or floor registers.

2. Basement odors and dust may be carried upstairs.

3. The basement will be warmed by the flow of heated air from upstairs, increasing fuel consumption.

4. If installed in an enclosed basement area without adequate make-up air, flue gases could be sucked out of the oil/gas furnace.

Before making this type of installation, check with your local building inspector as it may be prohibited by code.

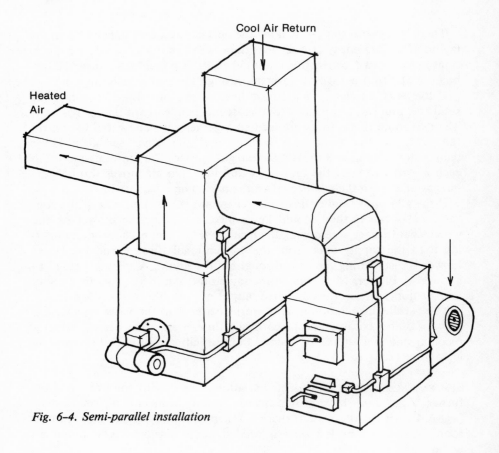

Fig. 6–4. Semi-parallel installation

ADDING ON TO A GRAVITY SYSTEM. There are several ways to make this connection. First, if the furnace is a converted wood or coal system, check with a heating contractor to see if it might be feasible to convert it back to solid fuel. You will have to remove the burner and firebrick combustion chamber and install grates and a new control system.

If this conversion can't be made easily, a separate wood or coal furnace may be the next best choice. The supply duct from the add-on should connect into the plenum above the existing furnace (fig. 6-5). Usually an add-on without a return duct is used, with the cool air supply coming from the basement stairs or floor registers.

DUCTS. When installing an add-on be sure that the existing ducts are large enough to carry the heat without overheating or causing excessive back pressure. In the case of a series system this can be ascertained by:

1. Calculation by a heating contractor.

2. Measuring temperatures and the blower motor amperage before and after the installation is made. The procedure includes operating the unit until it is warm (ten to fifteen minutes), then measuring the temperature in the plenum, and measuring the blower motor amperage.

After installing the add-on, the temperature of the plenum of the conventional furnace should be taken again. If it is more than 10° F. higher, the blower speed and/or motor size may have to be increased. The amperage should be checked again to see that it doesn't exceed the rating listed on the motor, otherwise the thermal overload may trip or the motor burn out.

In the case of a parallel installation, follow the manufacturer's recommendations for duct size. Check for large differences in air flow, using a monometer or air flow meter before and after the installation.

Duct connections should be tight, especially on the return air ducts, so heat losses are minimized and return air is not drawn from the basement.

The clearance from the ducts to combustibles should also be checked. Most conventional systems do not require as much clearance as a solid-fuel system. Refer to the installation manual or to figure 4-8.

Boilers

There are several ways to add a solid-fuel boiler to a conventional oil or gas boiler. Either series or parallel installations can be made. The series system is simpler but the parallel arrangement has some advantages.

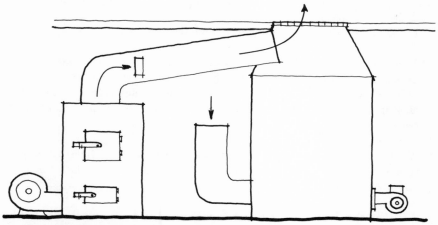

Fig. 6–5. Add-on to gravity furnace

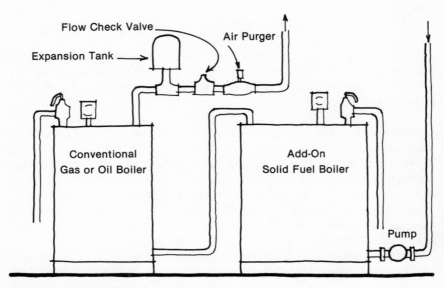

Fig. 6–6. Series installation

SERIES INSTALLATION. In this setup the return water from the radiators is pumped through the solid-fuel boiler and then through the conventional boiler before it is sent back to the radiators (fig. 6-6). If the wood/coal boiler is operating and heats the water enough, the conventional boiler will not start. The upstairs thermostat controls the circulating pump, the aquastat on the conventional boiler operates the burner, and the aquastat on the solid-fuel unit opens or closes the damper to keep the fire burning and the water hot.

Disadvantages to this arrangement include:

1. When the thermostat is not calling for heat and the pump is not operating, the water temperature in the solid-fuel unit where the fire is still burning may overheat, causing the temperature relief valve to open. This piping arrangement should never be used with boilers having a small water capacity.

2. The pipe layout should be arranged so that the return water goes through the pump before it is heated by the solid-fuel unit. Otherwise the pump life may be reduced by the hot water.

3. When the domestic hot water is connected to the conventional boiler, if water is drawn off it may start the burner even though the solid-fuel unit

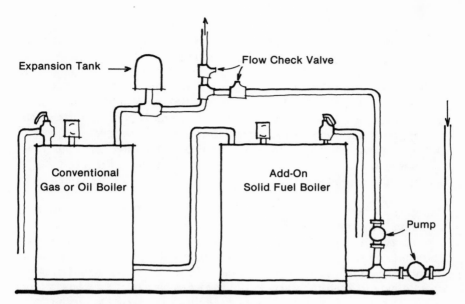

Fig. 6-7. *Series installation with feedback loop*

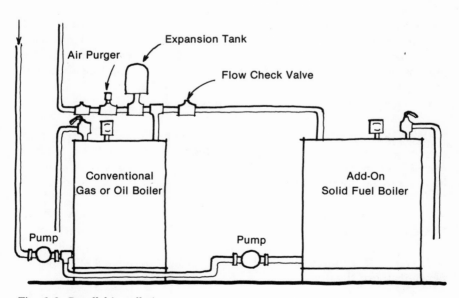

Fig. 6-8. *Parallel installation*

is operating. This can be overcome by installing a separate heat coil in the solid fuel unit that is piped into the water supply.

SERIES INSTALLATION WITH FEEDBACK LOOP. To overcome the disadvantages of the series system, a feedback loop with its own pump can be installed. As shown in figure 6-7, water from the solid-fuel unit is pumped through the conventional unit when the main circulating pump is not operating. A relay and check valves control the feedback loop pump and direct the water flow. Combining the capacity of the two water jackets reduces the possibility of overheating the solid-fuel unit.

PARALLEL INSTALLATION. This is the preferred although more expensive installation. Most manufacturers recommend this one.

In the typical parallel hookup, two circulating pumps are used (fig. 6-8). The pump on the conventional boiler is connected to circulate heated water when the room thermostat calls for heat. The solid-fuel boiler and its pump operate only to keep the water hot in the water jacket of the conventional boiler. The aquastat on the conventional boiler is connected to the burner so that it operates when the water temperature in the jacket falls below the set point. During the time of the year when the wood/coal boiler is operating and also for spring/fall operation, this setting can be 130° to 150° F.

The aquastat on the solid-fuel boiler is connected to operate the inter-boiler pump when the water jacket temperature exceeds the set point, usually at 160-180° F. Should the water in both units overheat, the limit switch on the solid-fuel unit will start the conventional boiler pump and dump heat into the house, cooling the water. The limit switch setting should be 190°–210° F.

The advantages to this type installation are:

1. There is less overheating because the volume of water is the sum of the two water jacket capacities. A solid-fuel boiler with a smaller water jacket can be used provided that the conventional boiler water jacket has adequate capacity to supply the radiators.

2. When the wood/coal unit is not being used, heat loss is reduced because water is not being circulated through it as in a series set-up.

3. With the addition of check valves, each unit can be operated independently of the other.

4. In the parallel arrangement, the solid-fuel boiler can be installed as a gravity unit by eliminating the interboiler pump and increasing the pipe size.

5. Operation during power outages is possible at a low fire rate if bypass loops around the circulating pumps are installed.

6. The domestic hot water coil in the conventional boiler will be heated from water circulated from the solid-fuel unit, saving energy.

MIXING VALVE. An additional improvement to an add-on installation is a mixing valve that provides a steadier demand on the boiler and quieter operation. The conventional boiler pump is wired to operate continuously. The mixing valve that is connected to the room thermostat is piped so that some of the returning cool water is mixed with the hot boiler water (fig. 6-9). The percentage of mix depends on the need for room heat. If more heat is needed, more hot water is added.

Although there is the added cost of operating the pump continuously ($20-30 per year), the savings by reducing the amount of overheating will likely offset this. The solid-fuel boiler will also be operating at a more uniform rate and creosote buildup may be reduced.

Fig. 6–9. Mixing valve

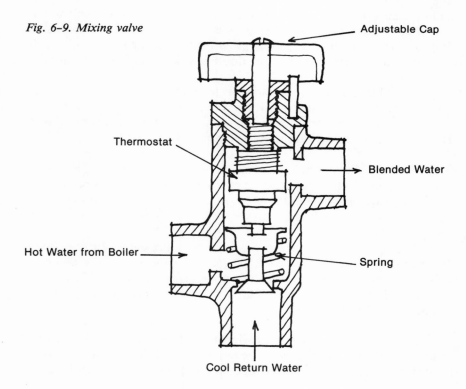

Adjustable Cap

Thermostat

Blended Water

Hot Water from Boiler

Spring

Cool Return Water

7 MULTIFUEL FURNACES & BOILERS

If you are building a new home or the furnace in your home is more than ten years old, you should consider installing a combination or multifuel system. Units are available that will use the conventional sources (fuel oil, natural gas, LP gas, or electricity) and will also burn solid fuels (wood, coal, or biomass). Pick almost any combination of these fuels and you can find a unit that will burn them.

Here are several advantages to this type of unit:

1. Probably the biggest one is that only one chimney flue is needed. Most homes do not have the extra Class "A" chimney that is needed by most add-on units. Finding a space for the extra chimney is often difficult and the cost of construction can make the installation hard to justify. Combination units are designed so that the flue gases from both the solid fuel and the conventional fuel leave the furnace through a single stovepipe connection.

2. A combination unit, although larger than a conventional furnace, takes up less than half the space of the conventional unit with an add-on. If you have a small furnace room or don't want to use up valuable space in the basement, the multifuel unit should be considered.

3. Most combination units use new high-efficiency burners. The oil and gas burners available today have an efficiency exceeding 80 percent. Your present furnace may be burning at 60 percent or less. The savings from installing a new unit even if you burn very little solid fuel will help to pay back the cost quickly.

For example, the savings from replacing your present oil furnace that is 60 percent efficient and burned 1,500 gallons last winter with a new combination oil/wood furnace that is 80 percent efficient will be 375 gallons or $562.50 at $1.50 per gallon even if no wood is used. If you burn some wood or coal, this savings will be much higher.

4. Installing a multifuel unit is easier and simpler than an add-on unit. If you are replacing your present boiler, very little modification may be needed to connect to the existing pipes or ducts. In a new installation, only

one connection is needed compared to multiple connections when a conventional furnace is used with an add-on. The controls are simpler to connect, being prewired and in place on the furnace. Usually only connections for a thermostat and power are needed. This saves on the installation cost.

5. Almost all multifuel units will switch from the solid-fuel burner to the conventional system if the fire dies down or goes out. This insures that your house stays warm should the fire burn out during the night or when you are away. In some states you cannot install a solid-fuel furnace or boiler without a backup. The combination unit provides this in one neat package.

6. The multifuel unit may be a good choice, knowing the uncertainty of fuel supplies. If you have a good supply of wood now but live on one or two acres, select a furnace that can be adapted to coal. Likewise, natural gas may be the best choice for the conventional fuel now but oil may be less expensive at a future date. Most units can be fitted with either a gas or oil burner. Check that your choice of unit can.

DESIGN

The combination furnace or boiler is designed to burn more than one fuel in single or separate fireboxes surrounded by a heat exchanger. A unit with a single firebox is sized large enough to take the wood or coal, and the conventional fuel burner, oil or gas, is attached through an opening in the side or endwall (fig. 7-1). The oil burner in a few units is located so that it can be used to ignite the solid fuel. This advantage is obvious in that you don't have to spend time getting the fire going. On the other hand, a burner located to do this is subjected to the flue gases, and the nozzle and electrodes can be clogged from the soot and creosote, creating more maintenance problems. The burner must also be protected from mechanical damage such as the wood hitting it.

Some units are built with separate fireboxes, one designed and sized for the conventional fuel, the other for wood or coal (fig. 7-2). You can expect some increase in efficiency, maybe 5 to 10 percent, in the conventional fuel section. This arrangement also reduces the potential for burner fouling from the wood smoke.

For information on what to look for in design of the wood or coal section of the furnace or boiler, refer to Chapters 3, 4, and 5. The unit should be designed with a firebox large enough to hold at least an eight-hour supply of solid fuel.

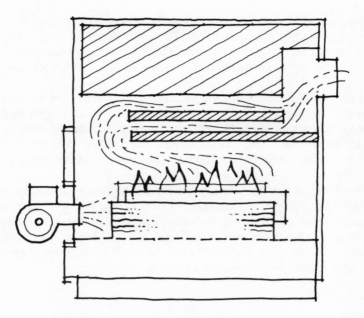

Fig. 7–1. This multifuel furnace has a single firebox.

Other design features to look for:

1. How is the unit shipped and assembled? Most units weigh from 700 to 1,000 pounds, and you have to get them into the basement. In some units shipping weight is reduced by removing the burner, blower, pump, and firebrick. This makes the unit easier to move but assembly time is increased.

2. How accessible is the unit for servicing? Heat exchangers, stove pipe connectors, burner nozzles, and filters need to be cleaned on a regular basis.

3. Can the unit be operated during a power failure? If so, what modifications have to be made?

FUELS

Your choice of conventional fuels generally depends on what is available in your area and its cost.

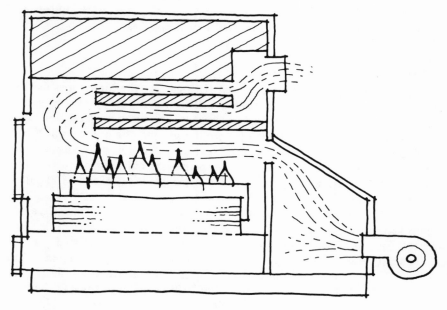

Fig. 7-2. This furnace has firebox for solid fuels, another for oil.

Oil

The high-speed flame retention burner is used by most manufacturers. This type of burner achieves greater than 80 percent efficiency. The combination of air-fuel mixture turbulence and less excess air creates a hotter, more concentrated fire. Most combustion chambers are made from a vacuum-cast alumina silica fiber combined with an inorganic binder. This material allows the temperature and efficiency to reach a maximum within a few seconds after the burner starts.

You will also need an oil tank and piping if they are not already installed. If you plan to burn mostly solid fuel, a 275-gallon tank is adequate. A 550- or 1000-gallon tank is a better choice if your wood supply is limited or you are away from home for extended periods. Oil dealers usually give a price break for larger deliveries, 500 or 1,000 gallons at a time. Oil will not deteriorate if stored for several years should you switch almost completely to wood or coal, but the tank should be kept full so moisture does not condense.

Be sure you purchase the correct grade of oil. Most burners used in multifuel units will burn the common No. 2 fuel oil. Some of the older units required the lighter, more volatile No. 1 or kerosene.

Gas

Manufactured, natural, and bottled (LP) gas are the three types used as heating fuels. Each of these gases has different combustion characteristics and heat values. Because of this, a gas burner is set up to burn only one type of gas fuel. A small percentage of a second gas may be mixed in at times.

Natural gas is currently the least expensive. Natural gas and manufactured gas are usually supplied by a pipeline system with a usage meter located at the entrance to your house. Bottled gas is supplied in 100-gallon replacable tanks or by truck to larger permanent tanks.

At the time of this writing multifuel furnaces with a gas burner were not approved, although several manufacturers offer this option for sale. The reason for this is that testing procedures have not been developed yet.

Power-type gas burners are generally used with spark ignition and safety controls. They operate similar to an oil burner in that the room thermostat controls operate the pump or blower and limit switches to prevent the unit from operating should a malfunction occur.

Electricity

If you plan to burn mostly solid fuel and need a backup for the occasional time you may be away for a day or two, electricity is a good choice. In new home construction this saves the cost of installing a fuel tank or connecting to the gas line. You will need to have an adequate supply of power at the distribution box, usually 100 amps minimum.

In furnace design the resistance heating element is located in the plenum or bonnet. The air heated by the heating element is carried through the duct system by air from the blower. The size of this element should match the needs of the home. Units are available from ten kilowatts (34,000 Btu per hour) to thirty kilowatts (102,000 Btu per hour).

Boilers have the electric heating elements, like those used in a hot-water heater, inserted into the water jacket. An aquastat controls the water temperature so that a set water temperature is maintained. Capacities from ten to more than thirty kilowatts are available.

A multifuel unit with an electric backup is less complex than one with conventional fuel, and therefore the cost is lower. Installation costs and maintenance problems are also reduced. Although the cost of electric heat is 20 to 50 percent more than oil or gas, you can often justify it because of the savings from the installation and maintenance costs if it is used only occasionally.

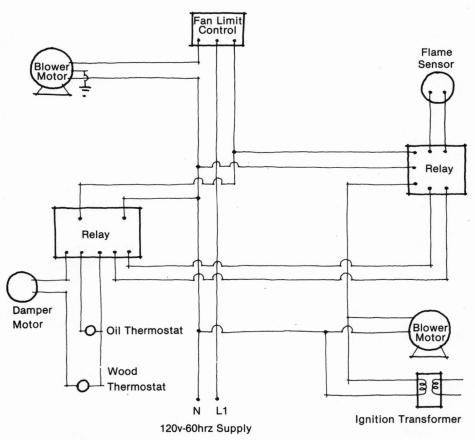

Fig. 7–3. This is a typical wiring diagram for a multifuel furnace.

Controls

The controls used on multifuel furnaces and boilers are similar to a conventional burner with an add-on solid-fuel unit with the exception that much of the wiring is combined into a neat prewired harness. State and local codes should be followed. Wiring drawings are included in most instruction manuals (fig. 7-3).

Here are several control features that you should look for in a multifuel unit:

1. Thermostat. Either one two-stage or two single-stage thermostats are used. When the unit is being operated with solid fuel, the setting on the con-

ventional fuel thermostat should be 3° to 5° F. below the desired room temperature. Should the wood or coal fire not be able to keep up with the heat demand or the fire die out, the oil or gas will take over.

One disadvantage that can occur with this, especially with natural draft, is that the solid-fuel fire may not pick up quickly enough, especially in cold weather. The oil or gas unit then kicks in for a short time. Some manufacturers include a delay timer that locks out the oil or gas burner for a period of a few minutes to an hour or more if you want to save on fuel. A greater spread between the setting of the two thermostats will achieve the same result.

2. A switch that shuts off the oil/gas burner should you open the fuel door is used by some manufacturers to prevent smell and fumes from being forced into the room. It is also a safety device should you forget to shut the fuel door.

Another switch that is often used prevents the burner from operating if the draft damper is open or the draft blower is on. This prevents the toxic oil or gas fumes from getting into the house.

3. Solid state (transistorized) controls are used by some manufacturers to control the burner. These are more precise and eliminate the contact points that have to be maintained and replaced from time to time.

4. A barometric damper (flap damper) or draft hood is needed in the stovepipe connector to control the draft on the burner. This should be set to the manufacturer's recommendations, usually between 0.03 and 0.06 inches water column. This setting should be adequate for a wood fire, but it may give problems when burning coal, which needs a strong draft. Inadequate draft will allow the fire to go out.

8 STOKERS

If you live in an area where coal, wood pellets, or wood chips are in plentiful supply at a reasonable cost you should consider a stoker-fed furnace. Although you may have to spend half again as much as the cost of the furnace or boiler, the increased efficiency and labor-saving convenience should offset this in a few years. Consider these advantages:

1. *Takes less effort.* A fire started in the fall will burn all winter provided there is fuel in the hopper and the ashes are removed.

2. *More efficient.* The fuel fed at a slow, continuous rate burns more efficiently than when fired intermittently by hand.

3. *Less pollution.* With continuous fuel feed and proper mixing of the fuel and air, combustion is more complete and less smoke is produced.

4. *Burns a less expensive grade of coal.* The mining and processing of coal produces more of the smaller pieces, and they tend to be less expensive than the nut and stove coal used in hand firing.

How Stokers Operate

Present designs were developed more than fifty years ago and, except for some refinements and a few modifications, are the same as they were then. Peak sales occurred during the 1930s with over 100,000 units being sold in 1937. Business dropped off from more than 300 manufacturers at one time until just before the 1973 energy crisis when only 8 or 10 remained. These old-line companies sold mostly industrial units and replacement parts for those stokers still in use. As demand has increased, these companies have resumed production, and several other manufacturers have pulled out old drawings and dusted off old patterns.

The two basic designs of domestic stokers available are overfeed and underfeed. Several other types are used only in industrial applications.

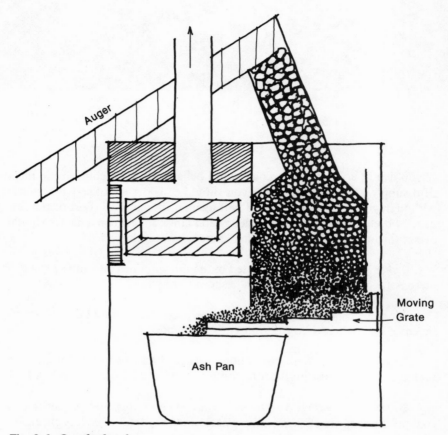

Fig. 8–1. Overfeed stoker

Overfeed

The simplest designs use an inclined or stepped grate located below a magazine that is kept full of fuel by augering from a bin or hopper (fig. 8-1). The weight of the fuel and the action of the fire create an ash layer on top of the grate. Powered by the same motor that operates the auger, the grate is moved back and forth an inch or two at a very slow rate. This cuts ash from the bottom of the fire bed and pushes it toward the ash pit. An induced draft fan in the chimney connector or draft blower supplying air below the grate is needed to provide adequate oxygen to the fire. The rate of burn is controlled by the room thermostat for a furnace or aquastat for a boiler, similar to conventional solid-fuel units.

Underfeed

In this type of stoker, the fuel is fed upward from underneath the furnace or boiler (fig. 8-2). The fuel is carried from a hopper or bin to the retort by the action of the screw. Variable rates of feed and therefore fire level are possible. The air for combustion is supplied from a blower and air tube to a wind box surrounding the retort. Besides the thermostat and limit switches, a hold-fire control is used to maintain the fire during mild weather.

The underfeed stoker is easier to use when converting a hand-fired furnace or boiler to an automated one. The firebox must be modified by removing the grate and ashpan and inserting the retort. Because correct installation is necessary to get proper operation, the unit should be installed according to the manufacturer's recommendations by a knowledgeable technician. Most units can be placed into the furnace or boiler through the side or rear, but room should be left to remove it for servicing or cleaning.

Stoker Construction

Although each company has its own variations in design, a review of the basic components will give you a better understanding of how stokers operate.

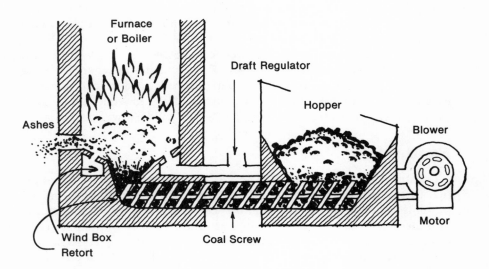

Fig. 8-2. Underfeed stoker

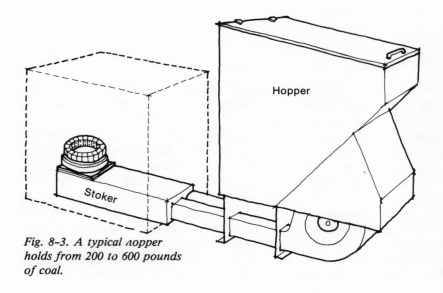

Fig. 8–3. A typical ʌopper holds from 200 to 600 pounds of coal.

Hopper or Bin

Most domestic stokers are available with a hopper designed to hold 200 to 600 pounds of fuel (fig. 8-3). This will provide enough heat for at least two days in cold weather and a week or more in mild weather. The hopper is made of sheet steel and is shaped to allow the fuel to flow freely to the screw. The steel is usually treated and protected to withstand the abrasion, chemical action, and moisture of the fuel. Most hoppers are airtight to keep the dust from getting into the basement.

A few manufacturers offer an auger feed system that will take the fuel directly from the storage bin (fig. 8-4). This eliminates the chore of filling the hopper. To operate properly, the bin must be located near the furnace, usually three to six feet away. It must also be designed with sloping sides, usually forty-five degrees minimum, so the fuel will flow without bridging.

Feed Screw

Located near the bottom of the hopper or bin, the feed screw carries the fuel to the furnace through a tube. Powered by an electric motor and transmission or ratchet mechanism, the screw turns at a very slow speed, in the range of one-half to six revolutions per hour. To allow variations in fuel feed, either a variable speed motor or multiple step pulley combinations may be used. With the step pulleys, the V-belt is slipped from one pulley ratio to another to change speed.

Although the feed screw has a lot of power because of the large reduction

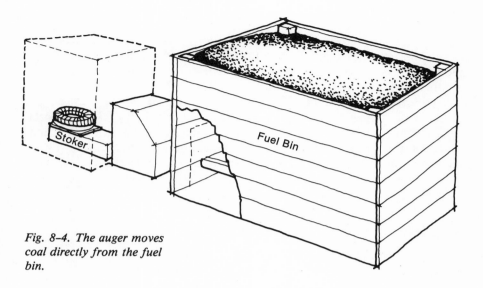

Fig. 8–4. The auger moves coal directly from the fuel bin.

in speed and can crush larger pieces of coal, some safety device must be incorporated should a piece of metal or other foreign object get into the fuel. This is usually done using either a shear pin that can be easily replaced or a slip clutch.

Feed screws are made of toughened steel to withstand the abrasion of the fuel. Two designs are common. With one, the screw turns within a metal tube and carries the fuel. With another, a metal band in the form of a spiral is welded to the inside of a rotating tube. This design, used in some overfeed systems, allows the fuel to fall back should the magazine be full.

Fan

The fan supplies forced draft to the fire in underfed units. It is usually attached to the hopper and powered by the motor that drives the feed screw. The air travels from the fan to the wind box surrounding the retort through an air tube or duct. The fan is equipped with either a manual or automatic control in the form of a damper to adjust the air supply. Depending on fan design, either a high or low pressure system is used. To prevent the passage of gases through the feed screw tube to the hopper, a small amount of air may be introduced from the fan to create a slight pressure.

Retort

The part of an underfeed stoker that supports the fire is called the retort. It is often shaped like the horn of an old Victrola phonograph. The fuel is forced up through the center by the action of the feed screw. Air from the wind box enters around the fire through air ports, called tuyeres (pronounced "two eyes"). Special designs may be used, depending on the type of fuel used.

The retort is an assembly of many small castings to allow for expansion because the temperature may reach 3,000° F. at full fire. Some retorts rotate to aid in ash removal. As the retort rotates, the ash is scraped off and falls into the ash pan or ash pit. The rotation also helps to remove clinkers and to break up coke trees. A coke tree is an accumulation of unburned coked fuel resulting from burning bituminous coal which becomes compacted into a vertical mass rising from the retort.

Grate

In the overfeed stoker, the fuel is supported by an inclined or stepped grate. When heat is needed, the induced draft fan, feeder auger, and reciprocating grate operate to increase the intensity of the fire. As ash builds up on the grate, a heat-sensitive switch activates the grate. Movement of the grate forces the ash into the ash container or ash pit. The fire settles down, and fresh fuel is added by the auger.

Two chambers are usually used with this system. The first, containing the grate, fuel chamber, and primary air supply, is where gasification of the fuel takes place. The second, integral with the first, is the combustion chamber where the gases are mixed with the secondary air and burned. These units are noted for their clean burning and high efficiency.

Controls

The basic controls and safety devices for a stoker-fired boiler or furnace are the same as for any unit that is not fired by a stoker. These should be reviewed in Chapters 4 and 5. In addition, two stoker controls may be used. The *hold-fire control* allows the stoker to operate at intervals during mild weather when the thermostat is not calling for heat but you do not want the fire to go out. The control usually contains two clocks, one that can be set for the length of time the stoker operates (½–10 minutes) and the second to select the time between cycles (30 or 60 minutes).

The *stack switch* is a thermostatically controlled switch placed in the first section of stovepipe from the furnace. If the fire dies out or the power is interrupted, this control will prevent the stoker from operating and filling the firebox with coal. A reset button must be pushed before the stoker will operate again. Some manufacturers replace the hold-fire control with a second type of stack thermostat that turns the stoker motor on when the

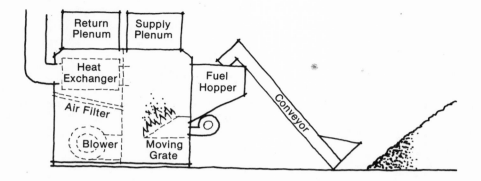

Fig. 8–5. This typical stoker-fired furnace burns sawdust, chips, or pellets.

temperature falls below a preset temperature. This feeds coal to the fire to keep it burning.

Fuels

Stokers require uniform size fuels to operate properly. With anthracite coal, usually No. 1 or 2 buckwheat size is used. This is available from most coal dealers. With bituminous coal, a low fusible stoker grade is generally recommended. Other factors that make good stoker coal include low ash content, freedom from fines, high heat content, and the use of a dustless treatment. The recommendations of the manufacturer and the local coal dealer should also be followed.

Three forms of wood are used with stokers. The best and most desirable is wood pellets. These ½- to 1-inch diameter by 1- to 1½-inch long pieces are made from sawdust or chips and when delivered are below 5 percent moisture content. They flow freely in bins and feed well with the auger. Because they hold their shape like coal when burned, the combustion proc-ess is uniform and complete. Other organic matter formed into pellets include straw, corn cobs, leaves, and animal manure. These all have about the same heat value, 8,000 Btu per pound.

Stokers are also available that will feed and burn sawdust and wood chips (fig. 8-5). Usually the overfeed method is used with a moving grate. Here uniformity of the pieces is important. The more uniform they are the fewer problems you will have. Usually a screen is placed on top of the feed hopper to remove large pieces and foreign objects. Moisture content is also impor-tant. Dry sawdust or chips feed and burn better than wet.

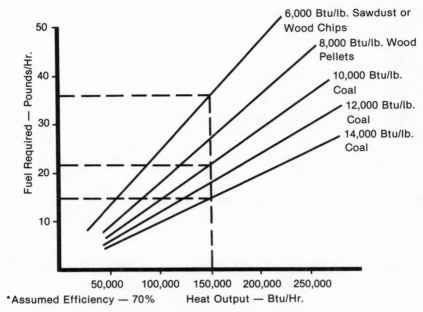

Fig. 8-6. In this example (dotted line), house requiring heat output of 150,000 Btu/hour will require about fifteen pounds of high-quality coal per hour.

Feed Rate

As can be see from figure 8-6, the fuel feed rate to keep your house warm depends on the type of fuel used and its heat value. This means that if you change from one fuel to another, the speed of the auger will have to be changed. For example, it takes about twice as many green wood chips as coal to give the same heat value.

Combustion is also important. Most furnace designs work best with one type of fuel, the one they were designed for. However, by adjusting the fuel feed rate, blower damper, grate speed, etc., other fuels usually can be burned. Check with the manufacturer or dealer before you purchase a unit.

Ash Removal

Burning a ton of coal will give from 50 to 200 pounds of ash. Provisions must be made to handle this amount, especially for stokers that may be checked only once a week. Most manufacturers provide a large galvanized ash container under the grate or retort. This will hold the ash from one hopper of fuel and should be dumped each time the hopper is refilled.

An automatic ash removal system is available from a few companies. This usually consists of an auger that carries the ashes from a pit under the retort to metal covered ash cans near the furnace. Automatic removal cannot be used if caking bituminous coal is burned.

Installation

If you purchase a stoker-fed furnace or boiler, the manufacturer's installation recommendations should be followed. Some companies ship the unit completely assembled and tested. All the installer has to do is connect the stovepipe, electricity, and piping or duct system.

When the stoker is to be installed in an existing unit, it's best to hire someone with experience to do it. There are minimum firebox dimensions, base heights, and setting heights that must be followed, and a tight seal is required. If these are not done, performance may be poor.

In cases where the furnace or boiler is not listed, the NFPA clearance requirements should be followed. For a furnace these require that the limit switch be set at no greater than 250° F. and that a barometric damper set no higher than 0.13 inches water static pressure be located in the stovepipe connector (fig. 8-7). For boilers that aren't listed, follow the NFPA recommendations shown in table 5-1.

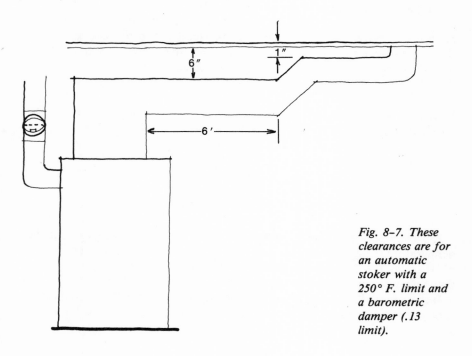

Fig. 8-7. These clearances are for an automatic stoker with a 250° F. limit and a barometric damper (.13 limit).

9 CHIMNEYS

In order to maintain combustion in the firebox, a continuous supply of air must move through the fuel bed. This air enters the furnace through the draft control openings and exits through the stovepipe and chimney.

Natural draft works much like a hot air balloon. As air is heated, it expands and becomes less dense. This allows the balloon to be buoyed up by the denser atmosphere. A chimney works on the same principle. The air containing the flue gases becomes lighter as it is heated in the firebox. It rises up the chimney. Cooler, denser air from the basement replaces it. This process continues taking heat and the flue gases away from the fire.

The intensity of the draft depends on the average difference between hot gases in the flue and the outside air, and upon the height of the chimney. The hotter the flue gas temperature, the greater the draft. For most chimneys, if the average temperature of the flue gases is kept 250° F. above the outdoor temperature there should be adequate draft.

To find the average flue gas temperature, use a candy or high temperature, probe type thermometer to take readings in the last section of stovepipe before the chimney and at the top of the chimney. A hole can be drilled in the stovepipe to insert the thermometer. Add the two readings, then divide by two. Then subtract the outside air temperature.

Example: the temperature at the thimble is 400° F. and at the top of the chimney, 300° F., with a 40° F. outdoor temperature.

$$\text{Temperature Difference} = \frac{400 + 300}{2} - 40$$

$$= 310° \text{ F. average flue gas temperature.}$$

A stovepipe thermometer placed on the pipe can be used to find the temperature being maintained, and, with a little experience, can be used to help control the fire. A few manufacturers provide a sensor that inserts into the stovepipe and connects to a gauge on the furnace.

Increasing the height of a chimney increases the draft and the amount of flue gases that can be removed. This is why furnaces in the basement

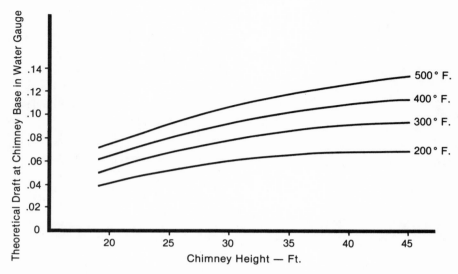

Fig. 9-1. To estimate draft, read up from chimney height to chimney temperature, then left to theoretical draft. The draft of a twenty-foot chimney at 200° F. is approximately 0.04.

generally burn better than those on an upper floor. The extra eight feet increases the draft by 0.03 inches at an average flue temperature of 300° F.

Measuring Draft

Draft is measured by using a monometer, a clear plastic or glass "U" tube partially filled with water. One end is inserted into the stovepipe or chimney and the other is left open to the atmosphere. The difference in the level of the water in the two sections of tube is the draft measured in inches. The draft in chimneys in most homes does not exceed 0.15 inches and is often near zero without a fire. Gauges have been developed to measure these low readings. Furnace servicemen and stove dealers often use these gauges when installing a furnace or if there are smoke problems.

The draft in a chimney can also be calculated. The curves in figure 9-1 give the theoretical draft at different chimney temperatures. Actual draft may be 10 to 20 percent less.

Practically all furnace and boiler manufacturers give the minimum required draft and/or chimney size and height in their installation manuals. These recommendations should be followed to give satisfactory performance.

The cross-sectional area of the chimney is also important. It must be large

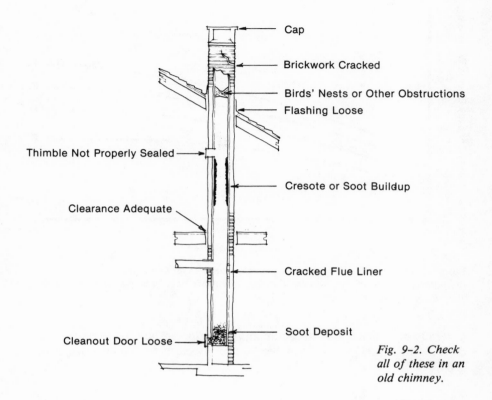

Cap

Brickwork Cracked

Birds' Nests or Other Obstructions

Flashing Loose

Thimble Not Properly Sealed

Cresote or Soot Buildup

Clearance Adequate

Cracked Flue Liner

Soot Deposit

Cleanout Door Loose

Fig. 9–2. Check all of these in an old chimney.

enough to carry the volume of flue gases and smoke generated by the fire. The minimum size should never be less than the size of the stovepipe collar on the stove. For example, if the manufacturer recommends an eight-inch stovepipe, a chimney at least eight inches in diameter should be used. Generally a slightly larger size is needed to overcome friction losses in the pipe and elbows.

Conversely, if the chimney is too large in area, draft and creosote problems can result. This is common in older houses where a large central chimney was built to serve several fireplaces. Some of these have inside dimensions that are six feet square. If you try to operate a furnace or boiler connected to one of these chimneys, you will generally have draft problems. The large surface area cools the flue gases too rapidly and the draft is lost. For best results the chimney cross-section area should not be more than 25 percent greater than the stovepipe collar. Table 9-1 can be used as a guide if there are no manufacturer's recommendations.

The draft must be great enough to overcome the resistance created by the fuel bed, the design of the unit, the stove pipe, and chimney. Otherwise the furnace will not operate properly. Fuel bed resistance is greater with coal or sawdust than with chunk wood, therefore a greater draft is needed. Heat exchanger designs that use two or three passes, or designs that have small fire tubes increase draft requirements. The same is true for long stovepipes or those having several elbows. By proper selection and installation of a unit, the deficiencies of a marginal chimney can often be overcome.

Forced Draft

To provide more positive control of the fire and aid in overcoming the resistance created by fuels like coal or sawdust, some manufacturers use a small blower to provide the primary air. These attach to the ash pit door or side of the furnace and blow air under the fire. This aids the natural draft in the chimney.

Induced Draft

In problem chimneys where the draft is not sufficient to keep the fire burning, an induced draft fan can be placed in the stovepipe connector. These have to be sized for the stovepipe and are available from stove shops and heating equipment companies.

Using an Old Chimney

Before connecting a furnace to an existing chimney, it should be thoroughly checked out, top to bottom. Older chimneys were not lined. Newer chimneys may be poorly built or have deteriorated. If you don't mind this type of work, including climbing on the roof, you might save yourself $40 to $50 but it's often best to get a local chimney sweep who has the necessary equipment and know-how to do the job. Let's look at what has to be checked (fig. 9-2).

● Is the chimney clean? You won't see much until you remove the soot, creosote, and birds' nests. A wire brush sized to fit the flue liner should be pulled up and down several times. The loosened soot and creosote should then be removed at the clean-out door at the base of the chimney.

● Chimney cap. The cap, if any, should be attached securely and be spaced at least three bricks' thickness above the top of the flue liner.

● The mortar that seals the bricks or stones to the flue liner at the top of the chimney should be sound and slope up toward the liner to help prevent down drafts.

● Flue liner. The tiles should be sound and not cracked. Mortar in the joints should not crumble when poked with a knife or awl. Give special attention to the mortar joints and the soundness of the bricks of chimneys without a liner.

● Chimneys with no flue liner or a cracked liner should be given a smoke test. Build a small, smoky fire, cover the top of the chimney, and see whether smoke is escaping anywhere. A factory-built chimney or heavy gauge stainless steel pipe can be used to line the chimney if the liner is unsafe. Be sure the joints are tight and the pipe is centered in the chimney.

● Clearance from wood beams. All beams and headers should be at least two inches from the chimney. In old chimneys where this clearance was not provided, sheet metal or aluminum foil slipped between the brick and the wood will give some protection.

● The clean-out door should be tight to prevent air from entering and reducing the draft.

TABLE 9-1. APPROXIMATE CHIMNEY CAPACITY

Chimney Flue Size (Nominal)	Capacity Btu/hr	Firing Rate*	
		Wood	Coal
6″ round	97,000	12.1 lbs/hr.	8.1 lbs/hr.
7″ round	131,000	16.3	10.9
8″ × 8″ square	150,000	18.75	12.5
8″ round	172,000	21.5	14.3
9″ round	218,000	27.3	18.2
8″ × 12″ rectangular	234,000	29.3	19.5
10″ round	270,000	33.0	22.5
12″ × 12″ square	358,000	44.8	29.8

*Assume 8,000 Btu/lb. wood
           12,000 Btu/lb. coal

Relining a Chimney

There are several methods that can be used to improve a chimney without a liner or one where the liner has been cracked. One method for a chimney without a liner is to have a mason insert sections of tile within the chimney and mortar these in place. It is best if the chimney is straight, without any

Fig. 9–3. A liner installed within a chimney

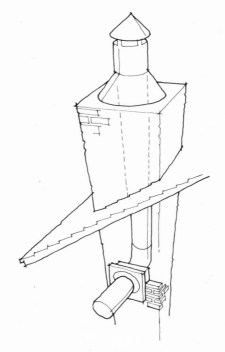

offsets. Sections of tile are lowered from the top of the chimney and set on top of one another. This job requires experience so that the tiles line up and very little mortar squeezes out of the joints. Only a few masons will attempt this type of installation.

A second method is to install a stainless steel liner or factory-built chimney within the existing chimney (fig. 9-3). An almost straight chimney is needed to use this method. Heavy gauge (twenty-four or heavier) stainless steel is used. Sections of pipe are connected with sheet metal screws or the conventional connecting methods for the factory-built chimneys and lowered into the chimney. Support must be provided at the bottom and spacers used every four to six feet to keep the liner from touching the brick. A thimble connection must also be provided for the stovepipe entering the liner from the furnace. At the top of the chimney a tight cap is needed to keep out rain and snow.

The Supaflu™ system*, a third method, was developed in Great Britain and uses a mixture that is poured around an inflated pneumatic tube inside the chimney. All cracks and bad mortar joints are filled. The inflated tube is

*National Supaflu Systems, Inc., Howe Caverns Rd., Cobleskill, NY 12043

sized for the diameter of the chimney needed. The mixture has good insulating characteristics that reduce creosote buildup. Although generally more expensive than a stainless steel liner, it will last longer and provide better service.

A New Chimney

Sometimes the best or only choice is a new chimney. Here you can select among masonry, faced with either brick or stone, concrete block, and factory-built. You will also have to select a location. A masonry chimney within the house acts as a heat sink, absorbing heat from the fire when it is hot and then giving it back when the fire goes out. External masonry and factory-built chimneys don't have this advantage.

When selecting an inside location, consider the space needed and look for obstructions. For an outside installation, check for location of windows, doors, and electrical wiring, and how the chimney will affect the appearance of the house.

If you plan to build the chimney, a good reference book to study is *Planning & Building Your Fireplace,* Garden Way Publishing Co. This gives diagrams, photos, and step-by-step construction details.

Effects of Flue Gases on Chimneys

Corrosion of chimneys and stovepipes can be hazardous. Flue gases emitted from wood and coal fires can form several types of acids including sulfuric and hydrochloric. When the acid vapors condense on the pipe and flue surfaces, usually when the temperature is between 150° and 300° F., these acids eat away at the metal and tile liners.

Thin gauge stovepipe may last only one to two years before rust spots and pin holes occur. That is the time to replace it. Tests have shown that some stainless steel pipe will corrode and rust after one year's use. Others have been in service for more than ten years with no apparent damage. Even tile liners may show some pitting and flaking after several years' exposure to these acids.

This deterioration can be slowed by maintaining a hotter fire so that condensation does not occur in the chimney system. Also when using coal, purchase one that has the lowest possible sulfur content (less than 0.6 percent if available). Finally, inspect and clean the chimney regularly.

Chimney Cleaning

This is a necessary chore associated with solid-fuel heating. With wood, creosote is formed from the unburned gases and condenses onto the cool stovepipe and chimney surfaces along with the moisture from the wood.

Creosote takes many forms. It can be thick and sticky like tar, black and watery like ink, or shiny and flaky like dust. Sometimes deposits build up rapidly. I've seen stovepipes blocked solid in three weeks. Other times a chimney may not have to be cleaned for several years. For this reason, it's best to set up a schedule to check your installation. Usually once a month is adequate.

The amount of creosote deposited depends on the density of smoke from the fire and the temperature of the flue pipes. In practice, the most creosote is generated during low, smoldering burns, at night or during mild weather. Many manufacturers recommend that the furnace or boiler not be operated when the temperature outside exceeds 40° F.

No Creosote in Coal

Coal doesn't create creosote because of the low volatile content of the fuel. However, it does deposit fly ash and soot in the stovepipe and chimney flue. Inspect them two or three times during the heating season and clean them at least once a year.

Chemical chimney cleaners are available at most stove shops. Recent tests indicate that their effectiveness varies. None has been shown to have any effect once the creosote or soot starts to build up. One method that seems to work well is to have a good hot fire for at least one-half hour each day. This can be dangerous if not done regularly, or begun when the chimney is not clean; chimney fires could result.

Chimney cleaning is best done with a stiff wire brush that is worked up and down the chimney flue or stovepipe. An industrial vacuum works well for cleaning up the mess. If you clean the chimney yourself, be careful when climbing on high, steep roofs. You may want to hire a professional chimney sweep who has the right size brushes, scrapers, and safety equipment.

Stovepipe

The connection between the furnace or boiler and the chimney is often the weakest link in the whole heating system. It is subject to high temperatures and corrosive materials. All joints must be tight to maintain chimney draft and prevent toxic gases from getting into the house. Improper stovepipe installations are probably the greatest cause of fires related to solid fuel heating.

The NFPA standard for six- to nine-inch pipe is twenty-four gauge or heavier sheet metal. Most stove shops stock the lighter twenty-six and twenty-eight gauge because it is cheaper. Although these materials are probably just as safe when new, they deteriorate more rapidly from oxidation and the acids developed, and need to be replaced every year or two. The heavier materials are stronger and will last several years.

Stovepipe is available in several forms: galvanized, blue steel oxide, and sheet metal painted with high-temperature paint or nickel-plated. The galvanized pipe is not recommended for use in living areas as it can give off toxic zinc fumes if overheated, at least until the zinc burns off. The temperature at which zinc melts is about 750° F. A corrosion-resistant pipe has recently been made available that is protected inside with a baked-on silicone synergistic coating and comes with a four-year warranty.

In planning your stovepipe installation, try to make it easy and convenient to remove for cleaning and reassembly. Pipe tees installed at strategic points can make cleaning and inspection much easier.

The following is a list of twelve installation pointers that will help you get a safe system:

1. Place the furnace so that the length of stovepipe needed to reach the thimble is as short as possible. Most manufacturers recommend three to six feet, and the code suggests not more than ten feet. This is more important with horizontal runs than with vertical ones. The pipe is a good radiator of heat, and the surface area of a six-foot length of six-inch diameter pipe is almost ten square feet, about one-third the heat exchanger area on some furnaces. This large surface can help to remove excess heat from the flue gases. Remember, though, in furnaces operated at low fire with the damper almost closed, it could result in rapid creosote buildup and reduced draft.

2. Keep the number of elbows to a minimum. Each elbow reduces the draft. Elbows are available in fixed and adjustable styles. With coal, either one is acceptable. With wood, the adjustable type may leak cresote.

3. Don't reduce the size of the stovepipe from the furnace to the chimney as this will reduce the draft. If the collar takes an eight-inch pipe, use eight-inch or larger, never seven-inch. If your unit was made in Europe and has a metric size collar, either use metric pipe or purchase an adaptor to the next size larger.

4. Stovepipe is not approved as a chimney and should not be used outdoors. Purchase a factory-built chimney instead.

5. Stovepipe should be visible throughout its entire length. For this reason it is illegal to install it through a ceiling or floor, or in a closet or enclosed wall. If the pipe has to run through an interior wall, a ventilated thimble or factory-built thimble should be used.

6. A bare metal stovepipe should not be placed closer than eighteen inches from combustible material such as a ceiling, wall, or curtain. If it

must be closer, use a heat shield. NFPA reduced clearance recommendations are shown in table 5-2. The same alternate covering materials as discussed under Clearance to Combustibles, Chapter 5, can be used here to dress it up.

7. Stovepipe is sold in 24-inch lengths with a useful length of approximately 22½ inches. Often it comes unassembled to save packing space. To assemble, fit the corrugations together and compress the pipe to reduce its diameter. Once assembled, squeeze the ends of the joint with a pair of pliers. Pipe can be cut to length with tin snips or a hacksaw.

8. Connect sections with three small sheet metal screws in each joint. Holes should also be drilled in the furnace collar for screws there. The screws keep the pipe rigid but still allow disassembly for cleaning. If you have a coal furnace, it doesn't matter which way the crimped end faces. If you will burn mainly wood, face the crimped ends toward the stove to allow any creosote formed to drain back into the furnace.

9. Stovepipe should never run downhill to the chimney or be installed lower than the stove collar; deadly carbon monoxide and other toxic fumes could escape into the room. It is better if it can be sloped up ½ inch or more per foot of length. If horizontal lengths of four feet or longer are used, the pipe should be supported with steel strapping or wire hung from the ceiling.

10. The use of a barometric damper (also called an automatic draft regulator or flap damper) is debatable. Some manufacturers strongly recommend its use, others are just as strongly against it. NFPA recommends its use for thermostatically controlled solid-fuel furnaces. This damper, which is common in oil furnace installations, removes fluctuations that normally occur in a chimney and maintains a uniform draft on the fire. It should be between the first two sections of stovepipe near the furnace. It is available for either horizontal or vertical pipe installation.

With most multifuel units the barometric damper is required to maintain even draft for the conventional fuel. For a wood-fire furnace one is generally recommended only if you have excessive draft, 0.5 inches water static pressure or greater. With this damper in place, room air is drawn into the chimney, cooling it, which can lead to greater creosote buildup. This type damper will also allow air in should you have a chimney fire, increasing its intensity and possibly cracking the flue. If you burn coal, this concern is not present and a barometric damper can be used safely.

11. Normally the joints between pipe sections are tight enough to operate the furnace safely. In cases where the draft measured in the first section of

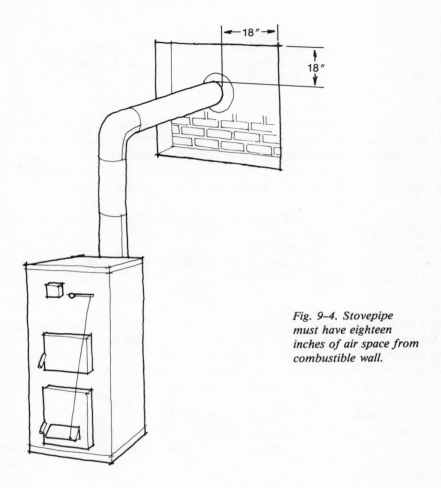

Fig. 9–4. Stovepipe must have eighteen inches of air space from combustible wall.

pipe near the furnace is marginal, applying furnace cement to the cor-
rugated end before assembly can increase the draft slightly.

12. Install the stovepipe so it enters the chimney horizontally. It should
be installed flush with and not extending into the flue lining. A tight connection should be formed between the stove pipe and thimble.

If the connection is to a masonry chimney, a combustible wall around the
thimble should be protected with eight inches of brick or an air space
eighteen inches from the pipe (fig. 9-4). The hole may be covered with noncombustible materials such as sheet metal or a UL listed wallboard. If the
connection is to an all-fuel factory-built thimble, only two inches of
clearance is required to combustibles. This space can be covered with a
stovepipe collar.

Dual Pipe Connections

Connecting a solid-fuel furnace or boiler to a flue being used by another heating appliance is allowed by some codes. The National Fire Protection Association Code adopted by many states does not allow this. Before attempting this, check with the local building inspector or fire marshal.

Some manufacturers specify in their installation manuals that their units should be connected only to a separate Class "A" chimney. There are some good reasons for this. They include:

1. The flue may be too small to vent the gases from both the solid-fuel unit and the conventional furnace, water heater, or stove. When both units operate at the same time the gases may back up, forcing fumes and odors into the house. The efficiency of both units may be decreased because of a reduction in the draft. If the chimney flue cross-sectional area is equal to the area of the larger stovepipe plus 50 percent of the second or if one unit is shut off while the other operates, the chimney flue may be large enough.

2. Flue gases from one unit may enter the house through the second unit. Carbon monoxide and sulfur dioxide are toxic. The barometric damper on an oil or gas furnace is the most likely spot because it does not close tightly.

A wood or coal furnace should not be connected to a flue serving a gas furnace having a draft hood in the stovepipe connector. This hood is designed to divert the back draft gusts of wind into the room before they reach the gas flame and blow it out. Because of this opening, wood or coal smoke could easily enter the house.

3. Deposits of soot and fly ash may build up in the bottom of the chimney, block one of the flue connections, and allow toxic gases to enter the room. If a dual-pipe installation is put in, the primary heating device connection, usually the oil or gas furnace, should be placed above the secondary device, usually the solid-fuel furnace.

4. Most oil furnaces leak a few drops of unburned fuel at times when valves do not close tightly. The air-gas mixture formed can be ignited by sparks or high temperatures from the wood or coal furnace and cause a puff back or small explosion.

On the other hand there are many stoves and furnaces that have been installed with the stovepipe vented into a flue serving another appliance. One dealer reports over 1,000 furnace and boiler installations without any problems; another, over 600. Fewer problems seem to occur with central systems because of higher flue temperatures and a greater draft. If at all possible avoid this and you will have a safer installation.

Heat Reclaimer

A heat reclaimer is a device that replaces a section of stovepipe and removes some of the heat that would escape up the chimney. The hot flue gases from the furnace or boiler pass through the housing and around the heat exchanger tubes. Air is blown through the tubes by the fan and picks up the heat, cooling the flue gases. The heat can be blown into the basement or ducted with stovepipe to other parts of the house.

With two exceptions heat exchangers should not be used with furnaces or boilers. They can extract too much heat, causing draft and creosote problems. The first exception is if the stovepipe is very short and doesn't radiate much heat. The second is with a poorly designed furnace or boiler that doesn't have enough heat exchanger area. Too much heat escapes up the chimney before it can be absorbed.

You can check your installation by using a high temperature thermometer (up to 1,000° F. — candy thermometer) inserted into a hole drilled in the stovepipe near the chimney thimble, or a surface-mounted stovepipe thermometer. If the temperature recorded during normal operation is above 600° F. you may be able to recover 10,000 to 20,000 Btu per hour without affecting the operation of your unit.

10 FIRING WITH WOOD AND COAL

Operating a furnace or boiler is a skill that is perfected over several years. That is not to say that you can't get a fire to burn the first time that you try, but that there are many intricacies with each fuel and with each heating unit that, when mastered, make firing easier and more efficient.

Before lighting your first fire in a new unit, give one final check to the installation. The best way to do this is to use one of the checklists in this book or to read through the installation manual. You want to make sure that you haven't forgotten something obvious, such as connecting the stovepipe to the thimble (this has happened). Also there are a few common sense "cautions" that should be reviewed.

1. Use the fuel or fuels that the unit was designed for. Don't burn trash, garbage, or batteries.

2. Never use charcoal lighter, gasoline, kerosene, or other highly flammable liquid to start your fire.

3. Don't light a fire if there are flammable vapors or dust in the area of the furnace.

4. Don't throw large pieces of paper such as Christmas wrapping paper onto a fire. It could set off a chimney fire.

5. Don't run the fire so hot that the metal turns red. This reduces the life of the furnace.

6. Don't operate a furnace with excessive draft. Follow the manufacturer's recommendations for draft requirements. Excessive draft can be reduced with a stovepipe damper.

7. Before fueling a furnace or boiler, open the primary draft for a few minutes to create a good draft and hot fire. Then open the door slowly.

8. Never leave the furnace unattended with the fuel door unlatched or the draft door wide open. This could lead to an overheated furnace or chimney fire.

9. Inspect stovepipe and flues regularly. Most chimney fires can be avoided if these are cleaned when necessary.

10. Don't dry or store fuel too close to your unit. Observe the recommended spacing of thirty-six inches.

11. Keep children away from hot furnaces. Temperatures exceeding 800° F. can be present.

12. Install a smoke detector near the furnace and also in the bedroom wing of the house. Many insurance companies reduce premiums if this is done.

Curing a New Unit

A new furnace or boiler should be fired lightly for the first week. Small fires allow castings to seat themselves and expand gradually. You may see some condensation on the furnace from moisture in the firebrick, especially if the unit was stored outside. Also, you will smell the fumes from the oils in the paints. These are normal occurrences and you shouldn't worry.

Small fires also give you a chance to get a feel for the unit and how it operates. Experiment with adjusting the controls, adding wood or coal, and cleaning the ashes. Check for water leaks from pipes, air leaks around ducts, and smoke leaks from flue pipes. On add-on and multifuel units observe if the conventional fuel burns when the solid-fuel fire dies down. As you build larger and hotter fires, adjust damper mechanisms so that the fire remains under control at all times. Verify that the settings on the safety controls, aquastats, and limit switches coincide with those recommended by the manufacturer.

BURNING WOOD

Learning to operate your furnace properly may come easily if you have had experience with a wood stove. With each unit being a little different in design and installation, you will have to develop your own technique. Follow the operation manual that the manufacturer has supplied. This section will review some of the basic principles and operating procedures.

Before starting the fire, be sure that there is power to the unit to operate the blower, pump, damper motor, and controls. This can be controlled by a switch located on or near the unit or at the power distribution box. Next, adjust the room thermostat to above the room temperature so that the dampers are adjusted as if heat were needed. On manual control units open the draft damper or ash door wide open. For add-on or multifuel units which use two thermostats, follow the manufacturer's instructions for the proper settings.

Be sure that the stovepipe damper and baffle bypass damper, if present, are open. This allows for direct full volume removal of the large amount of smoke generated when the fire is started.

If the unit has already been used, check the ash pan or ash pit to see that there are at least two inches of clearance below the grate for the air supply. Reduce the ash level on top of the grate to about one inch. A thin ash layer has some advantages in that it insulates the bottom of the fire and reflects the heat upward. Also, small pieces of charcoal left from the previous fire will provide heat for the new fire.

To start the fire, crumple up several pieces of newspaper and form a two- or three-inch layer over the bottom of the firebox. Magazine pages and colored newspaper should not be used as they do not burn well.

Add a layer of kindling, crisscrossing the pieces so that air and heat can get through. The best wood for kindling is dry softwood such as pine, spruce, or hemlock. The resins in them ignite rapidly and provide twice as much heat as the wood, giving a quick hot fire. Dead twigs, scrap lumber, and small pieces of pressed logs work well. The drier they are, the better. Try to keep a supply on hand under cover at all times. For kindling to ignite and burn rapidly it must be in small pieces, less than one inch in cross-section. Split sticks work better than round as more splinters are showing, offering a greater surface area. Remember the basics — to burn, the moisture has to be removed and the wood has to exceed the 550° F. ignition temperature. Before lighting, add three or four larger split pieces that will ignite as the kindling burns down.

Light the paper and close the door. This establishes the draft flow through the fire and out the chimney. It also keeps the heat within the fire box where it will start warming up the metal and firebrick. In a furnace or boiler weighing about a half a ton, it may take an hour or more to bring everything up to the normal operating temperature. As the unit warms up you may hear the crackling of the kindling burning and expansion of the metal parts, the piping, and the ducts.

Refueling

When the kindling is burning and a bed of coals is formed it is time to add a few more pieces of dry wood. Again, smaller pieces of split wood will catch more quickly. Leave about an inch of space between pieces so that draft paths are established and the heat from one piece will be reflected to the adjacent piece to keep up the temperature. A single large piece of wood will not burn well because the heat spreads out by conduction through the piece and is radiated away to be absorbed by the parts of the furnace, hence the wood surface temperature is not kept high enough to help the fire grow.

If your fire has failed, move the logs to one side of the fire box, add some more paper and kindling, and relight. They may catch this time as some drying and charring has occurred and they are already warm. Don't use any type of flammable liquid as the sparks and glowing embers could cause an explosion.

After the fire is burning brightly and you can start to feel some heat from the front, add another layer of wood. Until the furnace is fully warmed it is best to use smaller split pieces, up to four or five inches across. These ignite and come up to temperature quickly and start producing heat. Although some homeowners install a furnace rather than a stove so they don't have to split as much wood, the large pieces should be saved for periods of cold weather when the fire will be run hot, using a good draft, or for overnight operation when you want to carry a fire for a longer period of time. Larger chunks will burn well if they are placed to one side of the fire box and smaller pieces fed next to them. The smaller pieces burn hot, dry out the chunk, and eventually turn it to charcoal which then burns with even heat for an extended period.

On manual draft systems, you will have to keep an eye on the fire so you can reduce the draft once the unit gets up to temperature. Observe the water jacket temperature gauge and circulating pump operation on a boiler, and close the draft some as the house warms up. You should try to get a setting that will provide just the amount of heat that is being lost by the house. On furnaces, observe the blower operation and adjust the draft control so that the blower cycles on and off.

Automatic control systems will adjust themselves as the thermostat is satisfied. On units with damper motors, the draft door or damper is closed when the water in the jacket or air in the room reaches the set temperature. With a forced draft unit using a blower, the thermostat or aquastat turns off the power and the blower stops. In both cases some air enters and the fire continues to burn at a slow rate and is ready to be fired up again when heat is needed.

With a low fire for extended periods of time during the spring and fall, it's important that you use dry wood, otherwise creosote deposits may build up rapidly. Because of this, some manufacturers recommend that their units not be operated when the outside temperature is above 40° F.

Overfiring can be a problem. On manually controlled units that are left unattended overnight or during the day when no one is home, there is the potential for overheating should a rapid change in wind direction or speed occur. Although the safety devices such as the high temperature control are intended to maintain a safe temperature, they may not be enough, especially if the water jacket capacity is small or the air blower on the furnace is not adequate. A high temperature alarm such as the Chimney Fire Alert manufactured by Vermont Technology Group Inc., Morrisville, VT 05661, will offer some safety if someone is home. A better solution may be to subscribe to a telephone alarm system that alerts an answering service or fire department dispatcher. These are available from companies supplying security systems. Automatic controlled draft systems do not have as much chance of being overfired unless a door is mistakenly left open.

Ash Removal

The ashes should be shaken or removed as they build up. Too many ashes in the firebox will restrict the air flow to the fire and reduce the area for the wood. A layer an inch or two deep is desirable to support the fire and insulate the bottom of the firebox, but more than that should be removed. This may have to be done once every week or two if you burn wood.

Watch the smoke leaving the chimney, when you are firing with wood. In a properly operating furnace, most of what you see will be the water vapor which dissipates within a few feet of the chimney. If the smoke lingers on or trails across to the neighbors, your fire is not achieving complete combustion, and carbon particles and other chemicals are being emitted. The cure is to burn drier wood or operate the stove so that it maintains a higher firebox temperature.

BURNING COAL

The firing methods for coal are quite different than for wood. Because of the nature of the coal and the smaller amount of volatile matter, more primary air is needed. For example, on units designed to burn either fuel, draft settings may have to be adjusted to provide more air.

Different types of coal also require different firing methods. To light a fire with anthracite, you have to establish a hot bed of wood coals so that the temperature is above the ignition temperature, about 900° F. On the other hand, highly volatile bituminous coal is placed on the grate, and a small kindling wood fire is built on top.

Coal may be fed either automatically or by hand. Both methods have advantages and disadvantages. Hand-firing has been the traditional method and does not require the investment in expensive coal-handling equipment. Hand-firing requires more constant attention. It also is less efficient because of the more intermittent fueling and the frequent opening of the furnace doors. Stoker-firing will be covered in the next section.

Selecting the Right Coal

Use the size and type of coal that the manufacturer recommends. Generally, this is the type available in the area where the furnace was built.

If you have a used furnace or one that was made in another part of the country or world, what should you burn? You can with some experience burn almost any type of coal in any unit. There are a few features that differ and should be noted. The lower-ranked coals — lignite and sub-bituminous — have less heat value per pound. Furnaces designed for these fuels often have larger fireboxes than those designed for anthracite. More secondary air is needed for the lower-ranked fuels because of their greater amount of volatile matter.

The size of coal to use depends on the diameter or cross-section of the firebox, the depth to which the coal is fired, and the size of the grate openings. Larger sizes of coal can be used in larger fireboxes having a greater surface area and heat radiation. They don't pack as tightly together and so allow an easier air path. If you are having a problem with poor burning caused by insufficient draft, try using the next larger size coal.

The grate openings must be small enough to support the coal without allowing much to fall through to the ash pit. Generally, the largest grate opening dimension should be at least ¼ inch smaller than the minimum size of the coal.

The following recommendations were adopted many years ago by the anthracite industry.

Stove coal is suitable where the firebox is sixteen inches or more in diameter and twelve inches or greater in depth.* It is generally used in furnaces and only the largest of heaters.

Nut coal is the best choice for most stoves and smaller furnaces. It can be used where the firebox is up to twenty inches in diameter and ten to sixteen inches deep. It is also ideal for most kitchen ranges and fireplace grates.

Pea coal is used in smaller stoves, kitchen ranges, and other stoves having adequate draft. It is used in mild weather and in banking for the night.

Buckwheat is the smallest size that can be burned with natural draft. You must have an exceptionally strong draft to get satisfactory results. Buckwheat and the next smaller size, rice, find the widest use with stokers.

Other Coal Types

Bituminous coal and lignite are graded into many more sizes than anthracite, though generally only two or three sizes are available at the local

---

*Depth refers to the distance from the top of the grate to the bottom of the fire door.

coal yard. The coal dealer can help you decide which size will work best in your stove.

Recommendations for the size of *coke* to use are as follows:

No. 2: Nut — small stoves, kitchen ranges, hot water heaters

No. 1: Nut — larger stoves, heaters

Egg Size — furnaces and boilers.

All coal fires should be started with wood, to get the fire hot enough to ignite the coal. Charcoal, although easy to use and readily available, is expensive and may give off toxic fumes; it is not recommended. The wood should be dry. Softwoods make good kindling because of the resin they contain and because they split easily. Hardwoods are better on top of the kindling to give a fire that will last until the coal gets started. If you operate your furnace so that the fire goes out only a few times during the season, a car trunkful of kindling will be all you need.

BURNING ANTHRACITE

Anthracite is the most common coal used in the home. Its long, even burn time, high heat output, and cleanliness make it a good choice. Because it is used in industry, it is readily available in most parts of the United States. It is generally more expensive than other types.

Starting a Fire

Before starting the fire, open the smoke pipe, fire door, and ash pit dampers. Be sure that the power to the furnace is turned on and that the room thermostat is turned up so that it is calling for heat.

Place some crinkled newspaper and finely split kindling on the grate. Crisscross the kindling to allow air to get through. Use dry softwood. A few scraps of pine boards will get the fire started quickly. Light the paper. Add larger wood in a few minutes after the kindling is burning brightly.

Caution: never sprinkle gasoline, kerosene, or other explosive liquid over the kindling. This could cause serious injury.

Place the larger pieces on the fire so that they are slightly separated and form a level bed for the coal. It will take ten to twenty minutes before they are thoroughly ignited and ready for the coal. Adding the coal too soon can cut the air supply and smother the fire.

Add a thin layer of coal, preferably smaller chunks, to the wood fire, being careful not to disturb it too much or cut off the draft. Add a second, heavier layer when the coal is ignited and burning well. In furnaces with a shallow firebox, this may be all you will need to add until recharging. With deeper fireboxes, a third layer may be needed to bring the coal up to the bottom of the fire door. A red spot of glowing coals should be visible after firing. This insures that you haven't smothered the fire, and it helps ignite the gases given off by the new charge. A deep charge will give a more even heat and a longer fire. It may take one to two hours before the whole bed is fully ignited.

On a furnace with a manually controlled damper system, the dampers can be partially closed when the fire is well established and the house becomes warm. The secondary air supply, usually on the fire door, can be nearly closed. Leave the ash pit damper partially open; otherwise the fire will go out. Adjust the stovepipe damper to reduce the draft on the fire. With most types of anthracite you will see short, blue flames above the coal except when the fire is started or a new charge is added. If there is no flame, the fire needs more air from the bottom or it is near the end of its burn cycle and needs to be recharged.

On units with a thermostatically controlled draft, the damper will close automatically, reducing the intensity of the fire when the aquastat or room thermostat is satisfied.

Refueling

When the coal has burned down to half its original depth it's time to add more coal (fig. 10-1). Open the dampers and allow the fire to pick up a little and also burn off the gases. Open the fire door and pull the glowing coal to the front of the firebox with a coal hoe. Try not to disturb the fire too much. Add a fresh charge at the back, being careful not to seal off the top of the fire. Close the fire door but leave the fire door damper open for a few minutes until the volatile gases have burned off. It is not necessary to shake down the ashes each time you refuel most furnaces, but only experience can tell you what to do.

Shaking the Ashes

Be gentle when you shake the ashes. A few short shakes are better than a large movement of the grate. The objective is to remove a small amount of the ashes without disturbing the fire. The fire should be settled down about a half an inch or an inch in the firebox, until the first live coals start to fall. Excessive shaking wastes fuel by allowing unburned coal to drop into the ash pit. It can also expose the grate to very high temperatures, which can warp or burn it out. The fire may go out if you shake it too much.

With furnaces that do not have a grate that moves the entire bottom of the fire, you have to get in about every other day with a poker and push

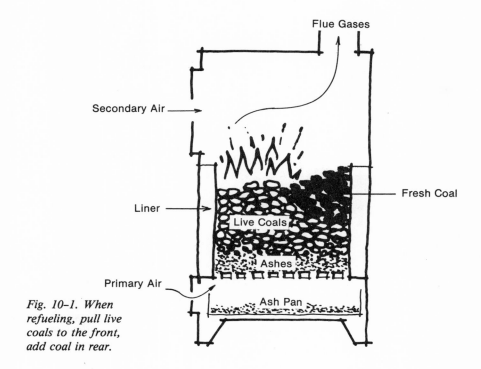

Fig. 10–1. When refueling, pull live coals to the front, add coal in rear.

down the ashes that build up in the corners. This is a problem with some of the units that are designed primarily for wood and have been modified with an optional grate system for coal.

Be careful in shaking the ashes so that you don't form clinkers. These form when very hot coal comes in contact with the ash layer. This occurs when you shake the fire too much or poke a fire. Some coals, especially those high in iron, form more clinkers. Because clinkers will not burn and will block the grate when formed in large pieces, remove them as necessary before refueling.

Banking the Fire

For overnight operation or if you are away all day, you will want to bank the fire. To do so, heap the coal up along the sides and back of the fuel box so that the fire gradually burns it over a longer period of time (fig. 10-2). You also reduce the intensity of the fire without letting it go out. Follow the same procedure as for refueling. If possible, avoid shaking, as a heavier layer of ash will help reduce the intensity of the fire during this time. In a furnace having poor draft control you can get longer burn cycles by using a

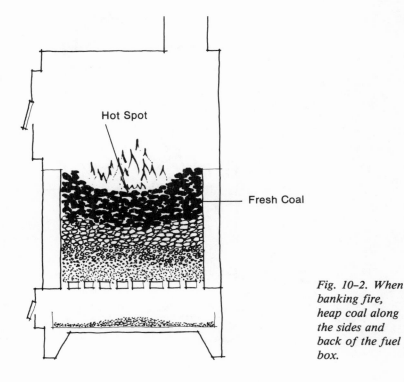

Fig. 10-2. When banking fire, heap coal along the sides and back of the fuel box.

layer of smaller pieces of coal, such as the fines that collect near the bottom of the pile, or a thin layer of ashes. This will reduce the combustion rate by impeding the flow of air.

After loading, let the fire establish itself for about a half-hour. This can be done by leaving the dampers open on a manual system or turning the room thermostat up higher on an automatic system. As you can see, it's important that the banking be done early enough before you retire or leave so that you can make adjustments after the fire is well established.

Reviving a Fire

Occasionally you may find that the fire is almost out before you remember to refuel it. You may first notice this as the house cools. First, open the ash door and stovepipe dampers and close the fire door damper to get a good draft through the grate. Then place a thin layer of dry coal from the top of the pile over the entire top of the fire. Do not poke or shake the fire. After the fresh coal has become well ignited, shake the grates and refuel.

After the Fire Goes Out

This will happen from time to time even to the most experienced fireman. You can, if you have dump grates, shake all the ash and coal into the ash pit, then screen out the coal for reuse. Often it is better to remove the coal through the fire door without disturbing the ash layer. Leaving an ash layer will protect the grates and help support the new fire. Establish the fire again by following the procedure under Starting a Fire.

BURNING COKE

Coke is similar to anthracite in its burning characteristics and most of the same procedures apply to both. A few differences need to be explained.

To make coke, coal — generally bituminous — is heated in an oven to remove the volatiles. As it is heated in the absence of air, it swells. The byproducts are recovered and used in many ways. What remains is mostly fixed carbon.

It is a relatively clean fuel. It may be dusty if it becomes too dry. A light sprinkling with a hose before handling will reduce the dust.

Because the volatiles have been removed, it is smokeless and doesn't create any soot accumulations. The chimney and stovepipe rarely have to be cleaned more than once a year.

Its porosity makes it a fuel that responds more quickly to changes in draft than anthracite, and it needs less attention between firings. The thermal efficiency of coke is equal to that of hard coal. Because it weighs half as much as anthracite by volume, more frequent refuelings are necessary. You can think of the relationship of coke to hard coal as being similar to the relationship of softwood to hardwood.

Because it doesn't burn as long, coke is less desirable if you want the fire to burn all night or all day without attention. A larger firebox is needed. You can also extend the burn cycle some by reducing the drafts, using a finer grade, or tamping the top of the fuel bed slightly to reduce the air spaces.

Getting a coke fire started can be a little more difficult. With most of the volatiles already driven off, a higher temperature is needed to get the carbon to burn. This means that you need a good wood fire first. Once the coke is burning, you can close the secondary air draft and control the fire with the primary air draft. Secondary air is not needed at this point. Adjustment of the dampers is more critical than with anthracite. With very few gasses or flames visible it is harder to see when a manual damper is closed too much. A surface thermometer placed on the stove or stovepipe may aid you in getting the proper adjustment.

In recharging, even in mild weather when you don't need a lot of heat, it is best to fill the firebox. This keeps a more uniform fuel temperature.

BURNING BITUMINOUS

Bituminous coal is very reasonable in cost in some areas, especially where it is mined, and therefore is a good choice, even though it is not as desirable a fuel as anthracite.

Because of its higher volatile content, bituminous is fired differently from anthracite. The *low volatile bituminous coals* — those with a volatile content of less than 20 percent — are generally fired with the conical method.

The first fire can be made similar to the anthracite fire. Use paper, kindling, and wood to get a bed of coals established. Add the coal in layers, allowing each to ignite before adding more.

This coal burns differently from the anthracite. Because it has more volatiles, the first flames will be long, generally orange or yellow. There will probably be quite a bit of smoke, too. As the gases burn off, the flames become shorter and may change color slightly, depending on the impurities present. The flame length also varies with the rate of burn, the longer flames indicating a hotter fire.

Once the fire is established, add the coal to the center of the firebox, forming a cone (fig. 10-3). The larger pieces will roll down the pile to the outside, allowing more primary air to flow through and creating a hot fire around the cone. This heat drives off the volatile gases, and the turbulence created increases the efficiency of the burn. After the volatiles are burned, the coke formed will burn more slowly, and you will get a long burn cycle.

Adjust the dampers about the same as for anthracite except allow more secondary air to enter, and open the stack damper until the volatiles are burned. Before refiring, break up the cone a little with a poker, especially if it has caked over or formed a crust. Be careful not to mix the coal. This increases the chances of forming clinkers. When shaking the grates, use short motions and stop when you see a glow in the ashes or the first red coals fall.

For overnight operation, shake the fire and add coal, forming the center cone. Allow the volatiles to burn off before closing the fire door and stack dampers or turning down the room thermostat.

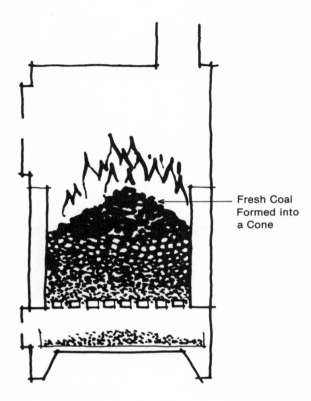

Fresh Coal
Formed into
a Cone

Fig. 10-3. Add a cone-shaped pile of coal when firing low-volatile bituminous coal.

You will have more maintenance with bituminous coal. In handling, there is more dust unless the coal was treated to remove it. Also, more soot will collect on heating surfaces and in pipes, requiring more frequent cleaning.

High-volatile bituminous coal, having more than 20 percent volatile content, is easier to burn but gives off more smoke. It burns somewhat like wood in that it is easier to ignite and burns with longer, smoky flames.

To start a fire, shake the ashes and remove the clinkers to clean the openings in the grates. Open the dampers. Shovel two or three inches of fresh coal onto the top of the grate. Crumple up a few sheets of newspaper and place them on top of the coal. Add a few pieces of dry, finely split kindling and a few larger pieces. Light the paper and close the doors on the furnace. When the coal is burning brightly, start adding coal in thin layers until the coal is up to the bottom of the fire door. Once the fire is well established, adjust the dampers, leaving the secondary air damper open a little so air can mix with the volatile gases.

An alternate method for starting the fire is to heap fresh coal against the back or one side of the firebox (fig. 10-4). Allow a little to cover the grate area. Place the newspaper and kindling against the sloping side of the coal

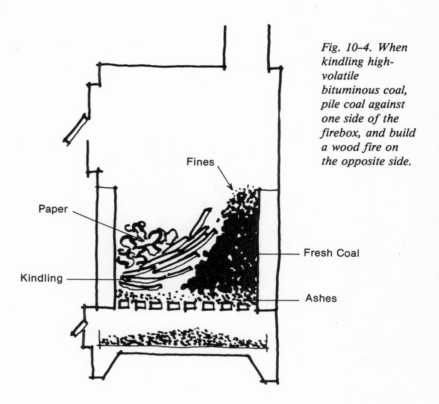

Fig. 10–4. When kindling high-volatile bituminous coal, pile coal against one side of the firebox, and build a wood fire on the opposite side.

and light the paper. This will ignite the pile from the outside and reduce the number of times that you have to add coal. When recharging, fill up the hole left where the kindling was, spreading the coal from side to side or front to back (fig. 10-5). Heat from the burning coal gradually penetrates the fresh coal, raises its temperature, and causes the gradual distillation of the volatiles. The hot coal causes a constant but slow burning of the combustibles and reduces the smoke to a minimum. It is important that any hot coals be shifted from the empty half before refueling, leaving just a layer of ashes. If not, you will have partial burning and a lot of smoke.

You can further reduce the smoke from this type of fire by covering the fresh charge with an inch or two of fine coal. Take this from the fines that accumulate at the bottom of the pile or purchase a few bags at the coal dealer. Adding the fines will keep the fire from spreading too rapidly and will force the gases into the hotter part of the fire. Before refiring, reduce the ash layer and break up the crust on the fuel side. The dampers are regulated the same as with other fuels except that more secondary air is

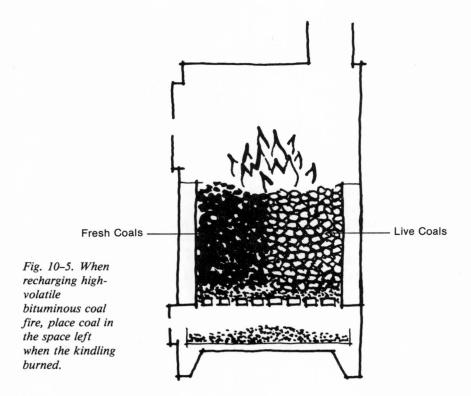

Fresh Coals ——

—— Live Coals

Fig. 10–5. When recharging high-volatile bituminous coal fire, place coal in the space left when the kindling burned.

allowed in to burn the volatiles coming off the top of the fire. More space should be left above the fuel when burning bituminous and other high-volatile coals to give a longer burning time before the gases reach the stovepipe. This will help to reduce the smoke.

BURNING SUBBITUMINOUS COAL AND LIGNITE

These low-rank fuels are not commonly burned in domestic furnaces or boilers except in areas near where they are mined. Subbituminous is found in Colorado, Wyoming, and Montana; lignite is found mainly in North Dakota and Texas. Their high moisture and ash content and low heat value make them less desirable than other types of coal. Overheating in storage can also be a problem. They should be stored in a tight bin with a cover.

Although standard furnaces and boilers will burn these fuels without difficulty, several units have been designed specifically for them. Certain firing practices will help to increase their efficiency. Deep fuel beds are not necessary because the coal burns at a much lower temperature. Once started, the fire does not die out as rapidly as it does with harder coals.

With these fuels a greater amount of fine fly ash is generated and blown around in the firebox. Clean all heat exchangers regularly, otherwise heat loss may result; also clean the stovepipe, or the draft may be reduced.

A fire with subbituminous or lignite will have to be tended more often; otherwise the firing procedures are about the same as with the hard coals. More secondary air is needed to get the highest efficiency but very little smoke is generated if a good fire with a hot surface is maintained. If you have, long, smoky, orange or yellow flames you are not getting fuel efficiency from the coal, and you should try to operate a hotter fire.

Firing Multifuel Units with Oil or Gas

Most multifuel and add-on units are fired the same way as a separate solid-fuel unit. The thermostat controlling the conventional fuel should be set 3° to 5° below the thermostat setting of the solid-fuel unit. This allows the wood or coal to carry the heat load before the conventional fuel unit starts.

One exception to this procedure is in the case of a multifuel unit that is designed so the oil or gas burner can be used to start the solid-fuel fire. With this type unit the firebox is loaded with wood and the solid-fuel thermostat is set about 5° F. above the indicated room temperature so that the draft damper or blower operates. The oil or gas burner thermostat is set 2° F. to 3° F. below the solid-fuel thermostat setting. This starts the burner, and once the wood fire is well established, the burner thermostat can be lowered so that it shuts off.

Procedures for Emergencies

With a safe installation and common-sense operation the occurrence of emergency situations will be infrequent — but still you should be prepared. An occasional review of the techniques that you should follow should you have a chimney fire or overheated furnace is recommended.

Chimney Fire

The loud roar of a chimney on fire can be frightening. Creosote will burn at temperatures in excess of 2,000° F. Quick action on your part can often prevent serious damage to your chimney and any combustibles that may be nearby. Memorize the following steps:

1. On furnaces and boilers with a room thermostat, turn the thermostat DOWN. This will close off the primary air supply to the fire. If you have an add-on or multifuel unit, turn both thermostats to a lower setting. With manual control units, shut the inlet dampers. Do not shut off the power to the unit as the blower or pumps will stop, causing overheating.

2. Alert everyone in the house.

3. Call the fire department.

4. If your unit or the conventional fuel unit that is being served by a common flue has a barometric draft regulator, use heavy aluminum foil or sheet metal to close the opening. Be careful that you don't get burned. Have some material handy that will do the job. Close secondary air supply dampers.

5. If you have a chimney flare, use it following the instructions. These give off large quantities of an extinguishing chemical that suffocates the fire. Don't breathe the extinguisher smoke, and inform the firemen when they arrive that you are using one. You may also try using a household type dry extinguisher with an ABC rating.

6. Be careful that you don't get trapped in the basement should a wall partition or other combustibles catch fire.

To reduce the amount of creosote that can form, use well-seasoned dry wood that has been stored under cover. Also split the wood into smaller pieces and where possible operate the unit with a hotter fire. Here is an example of how rapidly creosote can form. An anxious homeowner called me recently and indicated that creosote was leaking onto the floor from around the door of a box stove. A two-inch layer of the tarry form of creosote was found in the bottom of the firebox. The chimney had been cleaned and the stove installed only three days before. The cause of the problem: a new stove burning green wood at the lowest draft setting.

The best policy is to set up a maintenance schedule to check the stovepipe and chimney regularly, maybe once a week. You will have to determine the cleaning interval for your own furnace because each unit is installed and operated a little differently. Because creosote usually forms first in the stovepipe, you can use the tap test. Using your knuckle or a small hammer, tap the stovepipe. A hollow metal sound indicates it is clean, a dull sound indicates a creosote buildup. You should also disassemble a section of pipe and look inside occasionally.

Overheated Furnace or Boiler

If you forget and leave a door open or if one of the controls malfunctions, your unit could overheat. Normally the safety devices will try to compensate. Should you find this condition the following procedure should be followed:

1. Close the primary and secondary air dampers either manually or by lowering the thermostat setting.

2. Prop open the barometric draft regulator if there is one on your installation, to reduce the draft on the fire.

3. If necessary, cool the fire by dousing with small quantities of water. Use about a half-pint at a time and throw the water on the fire, not the hot furnace.

4. After the furnace cools, check the controls to be sure that they are still working.

Operation During a Power Interruption

Most furnaces and boilers can be operated to provide some heat during a power outage. Refer to the instruction manual that came with your unit.

Without power to operate blowers, pumps, and controls, heat circulation takes place slowly by gravity with the heated air or water rising and the cold air or water falling. Obstructions in the flow of heat have to be removed to make this as efficient as possible.

In hot air systems, the air filter is usually removed, duct dampers and registers opened, and the basement door left open for better circulation of the air. The access doors to the blower on some units are removed, on others they are not. On some hot water systems, the bypass valves around the pumps should be opened.

With both type units the size of the fire should be reduced because the heat cannot be transferred as rapidly. The air supply damper may have to be propped open a little, maybe ¼ to ½ inch. On units with bimetallic thermostats, turn the setting to low fire.

It is best when there is a power interruption to turn off the switch that controls power to the furnace. This will prevent damage to the motors should low voltage occur when the power is being restored.

When the power interruption is over, return the unit to its operating condition before the fire is increased and power turned on. Failure to do so could result in damage to the heat exchanger or controls.

STOKER OPERATION

Although a stoker-fired unit is more expensive to install, the advantages of achieving higher efficiency with less attention make it an attractive addition. The recommendations given here for firing a stoker unit are general in nature. You should refer to your instruction manual for details on specific settings and adjustments for your unit.

Underfeed Stokers

Before starting the fire, fill the hopper with fuel and remove any ashes. See that the coal feed and air damper on the blower are set properly for the time of year. A lower feed rate is used during the fall and spring when less heat is needed to keep the house warm.

On many units the feed rate is changed by releasing the tension on the drive belt and moving it to the proper combination of pulleys. Consult the manual for proper selection. Retighten the belt. Adjust the air damper so that more air is allowed for starting the fire.

Set the room thermostat above room temperature. Turn on the power and operate the stoker until the fuel appears in the retort, then stop it. Using paper and kindling, wood chips, or charcoal, start a fire on top of the retort if you are burning coal. When this is burning briskly, add a thin layer of coal on top. As soon as the coal is burning the stoker may be started. If you are burning wood chips, sawdust, or pellets, the crumpled paper may be enough to get the fire started.

The air supply should be adjusted after the fuel bed builds up. If the fire is low but unburned fuel is being dumped off the retort, increase the draft. If, on the other hand, the fire is hot but low on the retort, reduce the draft. The fuel should appear as a smooth mound with flames spread evenly over the top.

Anthracite flames are usually short, bluish, and smokeless. Bituminous coal burns with longer yellow to orange flames and, if properly adjusted, will show very little smoke. Wood burns with longer blue to orange flames that may contain more smoke and steam. With any fuel an intense, white fuel bed usually indicates too much air. A lazy, smoky fire probably needs more air.

The best way to adjust the air supply is to use an efficiency test kit. Most installers and servicemen carry these kits. The automatic air control used on most stokers is adjusted with the draft gauge to the setting recommended by the manufacturer. Flue gas temperature in the stovepipe and the carbon dioxide or oxygen level of the flue gases are measured several times during a half-hour period. The averaged readings are compared to a chart to determine overall efficiency. Slight adjustments to the air control can be tried to achieve a higher efficiency.

If the fuel feed rate is not enough to keep up with the heat needs of the house, as for example during a cold spell, the drive belt should be repositioned on the pulleys to achieve a faster feed. The air control will have to be readjusted to allow more air to the fire. In the spring or during warm spells, the rate may have to be reduced, otherwise the fire may die out. The hold-fire control, which allows the auger to feed fuel to keep the fire going even though no heat is being called for, can also be adjusted.

A stoker unit cannot be operated to provide heat during power outages unless a stand-by generator is used. A minimum 1,500-watt output generator is needed to operate most stokers. Care should be taken to connect the unit so that power cannot be fed back into the power lines where workmen may be trying to correct the problem.

Overfeed Stokers

With the overfeed stoker, a deep firebed is maintained on top of a moving grate. The auger feed keeps a constant level of fuel, and a primary air blower provides the draft.

Check the ash container and fuel bin before starting the fire. Leave a layer of ash on the grate if there has already been a fire in the furnace. Turn on power to the unit but do not start the feed auger.

Using paper and kindling, build a fire on top of the grate. Add a few larger pieces, and when the fire is well established and a bed of coals has formed, add a few shovels of fuel and turn on the air blower. Wait until this is burning brightly before adding another layer. After adding several more layers until the fire is well established to near the top of the firebox, the feed auger and grate motor can be started.

As with the underfeed stoker the air damper will have to be adjusted to match the rate of fuel feed and the natural draft. This can be done by observing the fire or by instruments. During cold weather the feed rate may have to be increased with a corresponding increase in the amount of draft air.

11 MAINTENANCE AND TROUBLESHOOTING

Your furnace or boiler should be kept in good repair at all times. This will result in better operation, greater efficiency, and a longer useful life. Proper maintenance will also insure that the unit remains safe to operate. The maintenance chores can be divided into those that need to be done daily, weekly, monthly, and at the end of the heating season. You should be able to perform most of these yourself. For those that you can't do, select a competent serviceman.

Daily

Remove the ashes daily or when they start to build up. Too much ash in the ash pit will restrict the draft so the fire does not burn properly. It could also damage the grate by causing it to overheat. If burning coal, remove clinkers.

Weekly

Clean soot and ashes from heat transfer surfaces when they start to build up. A 1/8-inch layer of soot can reduce the efficiency by as much as 10 percent. Use a wire brush or scraper or the tool provided by the manufacturer.

Pull cleaner on heat reclaimer to remove soot buildup.

Start oil or gas burner on multifuel units to insure that it is operational. The soot needs to be burned off and the electrodes dried out.

Monthly

Check stovepipe and chimney for soot and creosote. For the stovepipe, use the finger tap method or disconnect a section and inspect it. On the chimney, use a mirror and flashlight. Clean when the soot layer is greater than ¼ inch. A wire brush works well.

Inspect air filter and clean or replace if dirty. Reduced air movement decreases efficiency and could result in overheating.

Examine all doors and dampers to see that they fit tightly and are operating properly. Replace seals or adjust as needed.

Three Months

Oil motors, blower, pump, oil burner, gas burner, heat reclaimer. Use two or three drops of nondetergent motor oil (SAE No. 20) at both ends of motor.

Clean radiators, registers, and grilles. Dust and dirt buildup reduces heat transfer. Brush and vacuum the surfaces.

End of Heating Season

Adjust thermostats to lowest settings.

Turn off electric power to the unit.

Close oil or gas valve on multifuel units.

Disconnect the stovepipe and clean it. Clean the chimney, if necessary, then remove the soot and ashes from the clean-out at the bottom. If you don't have the equipment or don't want to climb on top of the roof, hire a chimney sweep. He has the equipment and know-how to do a good job.

Remove the ashes, soot, and clinkers from the fire chamber and ash pit. Clean all heat transfer surfaces. A wire brush and old vacuum cleaner are good tools for this job. Don't use your best vacuum cleaner, since this task may wear out the parts.

If your chimney doesn't have a cap, fit a piece of sheet metal so that it covers the top to keep out the rain. A stone or piece of steel suspended from the center of the cover inside the chimney will hold it in place.

Look for broken or warped grates, doors, or dampers and order replacement parts.

Check for cracks between castings where furnace cement or asbestos caulking has been loosened. Replace as necessary.

Coat the metal surfaces inside with a light oil or silicone spray for rust protection.

Place an open coffee can of silica gel inside the firebox to absorb moisture. This gel can be purchased at any drug store.

Inspect primary and secondary draft regulators for freedom of operation. Lubricate with light machine oil.

Check controls and safety devices. Clean electrical contacts and thermostat sensors.

Check operation of flow control, solenoid, and radiator valves.

Inspect piping for leaks.

Clean or change air filter.

Check the blower belt. Replace if necessary.

Stoker

At the end of the heating season a *stoker* should be prepared as follows: Remove the remaining fuel from the hopper and paint or oil the inside. Leave the cover open for air circulation.

Clean the retort and remove any ashes or clinkers from the burner. Clean and oil the electric motor and adjust the belts. Check the oil level in the transmission.

Oil the stoker screw. This can be done by removing the unit or by running heavily oiled coal or sawdust through the stoker and leaving the feed screw and tube full over the summer. This prevents corrosion and rusting.

Oil Burner

The oil burner on multifuel units should be cleaned and serviced before the heating season. This should be done by a competent serviceman and should include:

Changing the oil filter

Cleaning or replacing the nozzle and electrodes

Adjusting the fuel-to-air ratio for maximum efficiency

Checking for oil leaks

Cleaning the safety controls.

The oil supply tank should be filled to minimize condensation.

Gas Burner

The gas burner on multifuel units should be serviced every two to three years. If it is located within the solid-fuel firebox, it should be serviced each year. A competent serviceman will

1. Check the operation of the main gas valve, pressure regulator, and safety control valve

2. Clean the nozzle and other related parts

3. Run an efficiency test.

Troubleshooting

You can read this book and understand all about how a furnace or boiler operates, but you really don't learn how to fire the unit until you do it. Each installation is a little different, and only through experience do you become familiar with the proper technique for your own unit. Problems will arise with no obvious answers. Following are a few of the more common ones and some suggestions as to what might be the cause and remedy. Refer to the operator's manual for additional suggestions.

Can't Get the Fire Started

1. Kindling will not burn.
- Use finely split pieces of dry wood. Softwoods (pine, spruce, hemlock, cedar) work best because of the resins they contain. The wood should be dried at least a year and be stored under cover.

2. Chimney not drawing.
- Are thermostats adjusted properly so that the damper is open? If not, refer to operator's manual.
- Are the chimney and stovepipe dampers open?
- Is the chimney blocked?
- Heat chimney and get the draft started by lighting some newspaper on top of the kindling.
- Are ashes in the ash pit blocking the grate?
- If restarting a fire, no more than one inch of ashes should be left on top of the grate.

3. Wrong wood.
- Wood green or wet. Steam and water ooze out the end of the pieces. Obtain some dry wood and mix with green or wet wood.
- Pieces too large. Split to three- to four-inch diameter.
- The wrong firing method is being used. Refer to the section on firing.

4. Wrong coal.
- Use recommended type for your unit. Anthracite is harder to start than other types and requires a good bed of wood coals. Sometimes the use of a small amount of coal briquettes, usually a soft coal, helps to get the fire started more quickly.
- The coal is too large. It takes more heat to ignite larger pieces.
- The wrong firing method is being used. Refer to the section on Firing for the correct method for your coal.
- Too much coal at the start blocks the draft. Try adding a few pieces at a time.

Low Heat Output

1. Insufficient draft.
- Soot buildup in stovepipe or chimney.

- Air leaks around stovepipe, thimble, or clean-out door. Use furnace cement to seal.
- Chimney not tall enough. Minimum height should be fifteen feet. Twenty feet is better.
- Chimney too small. The 8×8 flue is the minimum size for most units; 8×12 is better.
- Chimney too large. Use a cone on top or a separate stainless steel pipe within the chimney.

2. Furnace not adjusted properly.
- Too much secondary air. A small amount of cool air can curb a fire.
- Primary air damper closed too much.
- Clinkers blocking grate.
- Ash buildup in corners of grate.
- Too much ash in pit.
- Green or wet wood. Use seasoned or dry wood for 20 to 30 percent more heat.
- Blower not delivering enough air. Check belt, blower speed, and air filter. Also check for obstructions over registers or radiators.
- Damper shutting off before thermostat is satisfied. Check fan and limit switches for malfunction. Check boiler water temperature.

3. Poor heat transfer.
- Soot on heat transfer surfaces inside unit. Clean with a brush.
  Poor heat distribution within the room or house. Use small fan, floor registers, or duct system.

4. Fuel.
- Not enough fuel in firebox.

Fuel or Smoke Smell in the House

1. Chimney related.
- Stovepipe damper closed. Open slightly.
- Chimney or stovepipe partially blocked with soot. Set up regular maintenance schedule to check this.
- Down drafts in chimney. Can be caused by nearby trees, a tall section of the house, or other building. Use chimney cap or extend the height of chimney one section.

2. Weather related.
- Temperature inversion with smoggy or foggy conditions resulting in poor draft. Operate with hotter fire.

3. Furnace related.
- Smoke entering house when fire door is open to refuel.
- Leaks in furnace or stovepipe connections.

- Puff backs when a mixture of combustibles, gases, and air is ignited from flame traveling through fuel bed.
- Burner malfunction. Contact qualified serviceman.
- Fuel leaks. Check fittings, valve seals, etc.
- Crack in heat exchanger. Contact qualified serviceman.

4. House related.
- House too tight. Bring in make-up air by opening basement door or installing separate pipe to outside.

Oil/Gas Burner Fails to Operate

1. Furnace related.
- No power. Check switch and fuse or breaker.
- Door safety switch open. Adjust switch.
- Fan and limit switch failure. Contact serviceman.
- Burner runs short time after pushing reset button due to worn or dirty components. Contact serviceman.

Coal in Ashes

1. Operator related.
- Too vigorous shaking. Shake only until first red coals drop into ash pit.
- Grate not positioned properly. Be sure there are no large cracks between grate sections.
- Coal too small. Use proper size for the unit.

2. Furnace related.
- Broken grate. Repair parts may be available through supplier or manufacturer.

Uses Too Much Fuel

1. Fuel related.
- Wet, green poor quality wood requires 20 to 30 percent more fuel. Use air-dried wood.
- Poor coal. Purchase coal with a low ash content from a reputable dealer.
- Wrong size. Use size recommended for the unit.

2. Furnace related.
- Leak in door due to worn or damaged gasket.

3. Operator related.
- Too much secondary air. Close the damper.

- Too much draft. Close stovepipe damper.
- Operating with too low a fire and not burning the volatile gases.
- Dirty air filters or radiators.

Pressure Relief Valve Trips Regularly

1. Installation related.
- Water jacket capacity too small. Add storage tank or connect to conventional boiler.
- Inadequate circulation. Add proper size circulating pump.
- Pipes installed incorrectly. Contact serviceman.

2. Operation related.
- Air in pipes. Bleed air from system or add air vents.
- Valve partially closed.
- Obstruction or restriction in the piping.

STOKER TROUBLESHOOTING

Stoker Will Not Run

- No power to motor. Check switch, fuse, breaker, or electrical connections.
- Auger overloaded or jammed. Press reset button. If still no operation, disconnect power and clean auger. Check shear pin.
- Limit control has shut off furnace. Allow to cool.
- Mechanical trouble. Call serviceman.

Excessive Coal in Firebox

- Too high a feed rate. Reduce auger speed.
- Clinkers blocking the retort. Clean the fire.
- Too little air. Blower duct partially filled with dust or damper out of adjustment.

Stoker Operates Continuously

- Control improperly adjusted. Call serviceman.
- Dirty fire. Let fire burn out, clean retort, and rebuild fire.
- Heat transfer surfaces dirty. Brush or vacuum.

Smoke Backs into Hopper

- Fuel level too low in hopper. Fill hopper.
- Clinker blocking retort. Remove.
- Fire burning down into retort. Too low rate of fuel feed or too little air.

Abnormal Noises

- Loose belt or pulley. Tighten or replace.
- From transmission. Low oil level or worn gears.
- From motor. Oil with two or three drops of SAE-30 motor oil.

CONCLUDING THOUGHTS

I hope that in this book I have pointed out the many factors that you should review when you consider a solid-fuel central heating system. Because it is a decision that will affect your family and their life-style for years to come, it should be made carefully. As a brief summary, the following questions should be answered.

Should I select a central heating system, or will a stove placed in the living area provide the necessary comfort?

Is wood or coal readily available at a cost that allows the new system to pay for itself in a few years?

Will my family put up with the inconvenience of maintaining the fire, added dust and dirt, and the occasional cold house?

Has the house been weatherized to reduce energy needs?

Which heating medium, water, air, or steam, is best for us?

Can I use an add-on to my present system to reduce the initial cost?

Should I consider a multifuel unit because no one is home all day to maintain the fire?

Will I take the time to develop the skills and techniques necessary to burn solid fuel properly?

How much labor will a stoker save me?

Are qualified installers and servicemen available locally?

Can the unit be installed to meet safety standards, codes, and insurance company requirements?

Will the operation of a solid-fuel furnace or boiler cause environmental concerns in the neighborhood?

As we progress through this era of uncertain conventional fuel supplies and prices, renewable resource solid fuels burned in central heating furnaces and boilers will be the answer to the heating needs of some of us. New designs and developments have made today's units simpler and safer to operate than those our grandparents used.

BIBLIOGRAPHY

AMERICAN SOCIETY OF HEATING, REFRIGERATION, AND AIR-CONDITIONING ENGINEERS, INC. *ASHRAE Handbook-1977.* New York: 1977.

BARTOK, JOHN W., JR. *Heating with Coal.* Charlotte, Vt.: Garden Way Publishing Co., 1980.

BRUMBOUGH, JAMES W. *Heating, Ventilating, and Air-Conditioning Library,* 3 vols. Indianapolis: Theodore Audel & Co., 1979.

EMERICK, ROBERT H. *Heating Design and Practice.* New York: McGraw-Hill Book Co., 1951.

JOHNSON, A.J., and AUTH, G.H. *Fuels and Combustion Handbook.* New York: McGraw-Hill Book Co., 1951.

MCGUINNESS, WILLIAM J., and STEIN, BENJAMIN. *Building Technology: Mechanical & Electrical Systems.* New York: John Wiley & Sons, 1977.

SHELTON, JAY W. *Wood Heat Safety.* Charlotte, Vt.: Garden Way Publishing Co., 1979.

STEINER, K. *Fuels and Fuel Burners.* New York: McGraw-Hill Book Co., 1946.

TWITCHELL, MARY. *Wood Energy: A Practical Guide to Heating with Wood.* Charlotte, Vt.: Garden Way Publishing Co., 1978.

CATALOG

MANUFACTURERS AND IMPORTERS

The following is a list of the addresses of manufacturers and importers of solid fuel furnaces, boilers, stokers and commercial size units described in this section.

ANTHRATHERM

Van Wert Manufacturing Co. Inc.
739 River St.
Peckville, PA 18452

ANTHRATUBE

Axeman-Anderson Co.
233 West St.
Williamsport, PA 17701

ASHLEY

Ashley Heater Company
1604 17th Ave. SW
Sheffield, AL 35660

BIRMINGHAM

Birmingham Stove & Range Co.
PO Box 2647
Birmingham, AL 35202

BIO-THERM

CHH Technology
PO Box 1387
Tappahannoch, VA 22560

BRUNCO

L.B. Brunk & Sons, Inc.
Salem, OH 44460

BUDERUS

R.W. Gorman
200 S. Washington Ave.
Washburn, WI 54891

CHARMASTER

Charmaster Products, Inc.
2308 Highway 2 West
Grand Rapids, MN 55744

COLUMBIA

Columbia Boiler Co. of Pottstown
Box G
Pottstown, PA 19464

COMBO

Combo Furnace Company
1707 West 4th St.
Grand Rapids, MN 55744

COMBUSTIONEER

Will-Burt Company
169 S. Main St.
Orrville, OH 44667

CONIFER

R.W. Gorman
200 S. Washington Ave.
Washburn, WI 54891

CONSECO

Conseco
611 North Rd.
Medford, WI 54451

COUNTRY COMFORT

Orrville Products, Inc.
375 E. Orr St.
Orrville, OH 44667

CROSSWINDS

Crosswinds Wood Energy Corp.
North Highway 63
West Plains, MO 65775

DAKA

Daka Corporation
Industrial Park
Pine City, MN 55063

DANIELS

Sam Daniels Co., Inc.
Box 868
Montpelier, VT 05602

DAKOTA

Western Heating, Inc.
PO Box 1382
Dickinson, ND 58601

DEDIETRICH

Heatek
45 W. 55th St.
New York, NY 10019

DEFIANCE

Defiance Company
Calumet, MI 49913

DORNBACK

Dornback Furnace & Foundry Co.
33220 Lakeland Blvd.
Eastlake, OH 44094

DOVER

Dover Stove Company
Sangerville, ME 04479

DUAL-AIRE

St. Croix Wood, Wind & Sun
Main Street
Roberts, WI 54023

DUO-MATIC

Duo-Matic Olsen, Inc.
2510 Bond St.
Park Forest South, IL 60466

DUMONT

Dumont Industries
Main Street
Monmouth, Me 04259

DUAL-PAK

Robert Bell Industries, Ltd.
PO Box 70
Seaforth, Ontario N0K 1W0

EMF

EMF
Divison of General Machine Corp.
Emmaus, PA 18049

ENERGEN

Energen
PO Box 11216
Newington, CT 06111

ENERGY KING

Chippewa Welding, Inc.
Rt. 5, Box 190
Chippewa Falls, WI 54729

ENERGY-MATE

Energy Research & Development, Corp.
Town Line Road
Tomah, WI 54660

ESHLAND

Eshland Enterprises, Inc.
PO Box 233
Greencastle, PA 17225

ESSEX

Essex Thermodynamics, Inc.
PO Box 817
Essex, Ct 06426

FIRECRAFT

Ford Products, Corp.
Ford Products Road
Valley Cottage, NY 10982

FRANCO BELGE

Franco Belge Foundries of America
15 Columbus Circle
New York, NY 10023

FRANK'S PIPING

Frank's Piping Co., Ltd.
PO Box 160
Bromptonville, Que. J0B 1H0

FROLING

Froling GmbH & Co.
Kessel-Apparatebau
PO Box 5140
D-5063 Overath Bez. Koln, W. Germany

FURNACE WORKS

The Furnace Works
Star Route, Box 234
Babbitt, MN 55706

G & S MILL

G & S Mill
75 Otis St.
Northboro, MA 01532

GYRO

S & B Marketing, Inc.
14025 NW 58th Court
Miami Lakes, FL 33014

HARDIN

Hardin Manufacturing Co.
3956 Highway 119
Longmont, CO 80501

HOGFORS

Kimi Kymmene Engineering
SF-03600 Karkkila, Finland

HOT SHOT

River City Furnace LTD
2009 4th St. SW
Mason City, IA 50401

HOVAL

Arotek Corporation
1703 E. Main St.
Torrington, CT 06790

HUNTER

Woodburning Specialties
PO Box 5
North Mansfield, MA 02059

HYDRAFIRE

Hytek Systems, Inc.
360R Otis St.
Mansfield, MA 02048

JENSEN

Jensen Metal Products, Inc.
820 Water St.
Racine, WI 53403

JOHNSON

Johnson Energy Systems, Inc.
7350 N. 76th St.
Milwaukee, WI 53223

KERR

Kerr Controls, Ltd.
9 Circus Time Rd.
South Portland, ME 04106

KEYSTONER

Kogen Industries, Inc.
PO Box 370
Carrolltown, PA 15722

KICKAPOO

Kickapoo Stove Works, Ltd.
Box 127, Main Street
La Farge, WI 54639

KINETIC

Kinetic Industries, Inc.
122 Weiler Rd.
Elk Grove, IL 60005

KODIAK

Alaska Company, Inc.
480 W. 5th St.
Bloomsburg, PA 17815

KOPO

Kopo, Inc.
2000 Rhode Island Ave. N.
Minneapolis, MN 55427

LEADERS

Leaders Heat Products
11219 Red Arrow Hwy.
Mattawan, MI 49071

LONGWOOD

Longwood Furnace Corp.
Rt 1, PO Box 223
Gallatin, MO 64640

LYNNDALE

Dynndale International, Inc.
PO Box 1154
Harrison, AR 72601

MASCOT

Hiestand Distributors
Rt 1 Box 96
Marietta, PA 17547

MEMCO

Melvin Manufacturing Corp.
Rt 133 & 156
Jay, ME 04239

MENOMINEE

Menominee Boiler Works
1824 10th Ave.
Menominee, MI 49858

MONARCH

Monarch Appliance Division
Malleable Iron Range Company
715 N. Spring St.
Beaver Dam, WI 53916

NATIONAL STOVE WORKS

National Stove Works
Howe Cavern Road
Cobleskill, NY 12043

NECA

Arrow Heating Equipment Ltd.
1703 E. Main St.
Torrington, CT 06790

NEW ENGLAND BOILER

New England Boiler Mfg., Inc.
790 Old Colony Rd.
Meriden, CT 06450

NEWMAC

Newmac Manufacturing, Inc.
PO Box 545
Woodstock, Ont. N4S 7Y5

NORTHLAND

Northland Boiler Company
Town Street
East Haddam, CT 06423

ONEIDA ROYAL

Oneida Heater Co., Inc.
Box 148, 109 N. Warner St.
Oneida, NY 13421

PASSAT

Passat U.S.A. Inc.
1 North Rd.
East Kingston, NH 03827

PRAIRIE

Prairie Wood Boilers
Prairie Farm, WI 54853

PRIBBS

Pribbs Manufacturing
Highway 220, North Rt. 22
East Grand Forks, MN 56721

PRILL

Prill Manufacturing Corp.
PO Drawer S
Sheriden, WY 82801

QUAKER

Quaker Stove Co., Inc.
200 W. 5th St.
Lansdale, PA 19446

RITEWAY

Riteway Manufacturing Co., Inc.
PO Box 99
Weyers Cave, VA 24486

RUSSELL

Decton Iron Works, Inc.
21385 W. Good Hope Rd.
Lannon, WI 53046

SEVERANCE

The Energy Forum
672 Pleasant St.
Norwood, Ma 02062

SEVILLE

Earth Power
486 Boston Post Rd.
Orange, CT 06477

SHENANDOAH

Shenandoah Mfg. Co., Inc.
PO Box 839
Harrisonburg, VA 22801

SIMPLEX-MULTITHERM

Van Wert Mfg. Co., Inc.
739 River St.
Peckville, PA 18452

SPARTAN

ICR Associates
Trumbull, CT 06611

STADLER

Stadler Corporation
PO Box 180
Carlisle, MA 01741

STEEL KING

Steel King Industries, Inc.
325 E. Beckert Rd.
New London, WI 54961

SUREFIRE

Energy Options, Inc.
7408 Vine St.
Cincinnati, OH 45216

STOKERMATIC

Stokermatic
1610 Industrial Rd.
Salt Lake City, UT 84104

SUMMERAIRE

Americal Energy Marketing Assoc. Inc.
4642 Crossroads Park Dr.
Liverpool, NY 13088

TARM

Tekton Corporation
Rt 116
Conway, MA 01341

TASSO

ITS, Inc.
PO Box 900
Exeter, NH 03833

THERMO CONTROL

National Stove Works
Howe Caverns Road
Cobleskill, NY 12043

THERMO PRIDE

Thermo Products, Inc.
Highway 119
North Judson, IN 46366

TIMBER-EZE

Timber-eze, Inc.
Rt 5
Millersburg, OH 44654

TRITSCHLER

Covinter, Inc.
70 Pine St.
New York, NY 10005

XX CENTURY

Twentith Century Heating & Vent.
96 Ira Ave.
Akron, OH 44301

UMCO

Union Manufacturing Co., Inc.
PO Box 522
Boyertown, PA 19512

UNIVERSAL ENERGY RESOURCES

Universal Energy Resources
N. Parker Place, Suite 164
Aurora, CO 80014

VALLEY FORGE

Valley Forge Stove Co.
85 E. Bridge St.
Spring City, PA 19475

WEBER

ORC Industries
2700 Commerce St.
La Crosse, WI 54601

WILL-BURT

Will-Burt Company
169 S. Main St.
Orrville, OH 44667

WOODBINE

Woodbine Corporation
2919 Industrial Park Dr.
Finksburg, MD 21048

WOODCHUCK

Woodchuck Furnace Co.
119 S. Dewey St.
Eau Claire, WI 54701

WOODMASTER

Suburban Manufacturing Co.
PO Box 399
Dayton, TN 37321

YUKON

Yukon Industries, Inc.
9890 NE Highway 65
Minneapolis, MN 55434

SOLID-FUEL FURNACES

The following is a listing of a variety of styles and types of solid-fuel furnaces. Most of these can be installed as an add-on to a conventional furnace system.

Manufacturer and Model	Dimensions Height by width by depth	Blower Capacity Cu.ft./Min.	Weight pounds	Heating Capacity Btu/hour
Ashley				
F-78-80	37 × 21.1 × 48	700	321	60,000
Charmaster				
Wood/coal	54 × 28 × 55	1780	646	na
Country Comfort				
CC 2025(wood)	44 × 24.5 × 31.5	865/1065	480	na
CC 2035(coal)	44 × 24.5 × 31.5	865/1065	480	na
Crosswinds				
2030	50 × 31 × 34	1400	850	75,000
2430	60 × 32 × 36	2500	950	120,000
Daniels				
RCD–28	56 × 36 × 32	na	na	100,000
RCD–30	56 × 36 × 41	na	na	130,000
RCD–38	56 × 36 × 47	na	na	145,000
RCD–42	56 × 36 × 53	na	na	165,000

Manufacturer and Model	Dimensions Height by width by depth	Blower Capacity Cu.ft./Min.	Weight pounds	Heating Capacity Btu/hour
Daniels				
R30W	52 × 36 × 41	na	na	100,000
R36W	52 × 36 × 47	na	na	128,000
R42W	52 × 36 × 53	na	na	157,000
Dornback				
CW100–17BP	48 × 39 × 25.3	na	355	100,000
Duo-matic				
CWF	52.6 × 48.8 × 33.8	1000	625	120,000
Energy King				
110 M	35 × 24 × 29	500/750/1000	440	100,000
150 M	48 × 24 × 33	500/750/1000	575	150,000
Hot Shot				
RC 1900	39 × 22 × 41	465	378	110,000
RC 2400	44 × 24 × 41	930	438	150,000
Johnson Energy Systems				
J–6600	40 × 24 × 32	930	na	160,000
J–8800	40 × 24 × 32	930	na	160,000
Kerr Controls				
Scotty	43 × 22.5 × 46.5	1020/1200	344	70,000
Mini Scotsman	44.5 × 24.4 × 31.5	na	286	80,000
Scotsman	44.5 × 29 × 39	na	392	140,000
Keystoner				
Wood 100	36 × 22 × 28	465	400	100,000
Wood 120	47 × 24 × 30	865	500	120,000
Coal/wood 170	50 × 24 × 30	865	800	180,000
Kickapoo				
BBR–D	35.5 × 26 × 46	1075	na	75,000
Kodiak				
	41 × 24.5 × 51	1075	na	140,000
Newmac				
Wood WB 100	42 × 26 × 54.3	1200	600	100,000
Wood/coal				
WG 100	46 × 26 × 57.3	1200	650	100,000
Oneida Royal				
S20	52 × 40 dia.	gravity	450	72,000
22C	65.5 × 44 dia.	gravity	590	99,000
24C	66.5 × 48 dia.	gravity	790	127,000
27C	66.5 × 52 dia.	gravity	925	144,000
22JC	56 × 31.5 × 37.1	gravity	625	99,000
24JC	56 × 39.3 × 43	gravity	825	127,000
Passat				
HOL–20	55 × 29.5 × 72.8	na	na	88,000
Pribbs				
Wood Burner	64 × 30 × 68	1500	680	75,000
Riteway				
LF–30	65 × 36 × 73	na	1800	160,000
Russell				
	42 × 22.5 × 28	800	600	70,000

Manufacturer and Model	Dimensions Height by width by depth	Blower Capacity Cu.ft./Min.	Weight pounds	Heating Capacity Btu/hour
Seville				
FA 80	46 × 22 × 30	500/1000	410	120,000
Shenandoah				
Wood F–77	44.5 × 24 × 32	900	400	75,000
Coal F–77C	44.5 × 24 × 32	900	423	75,000
Spartan				
F740S	47 × 23 × 32	525	508	na
Steel King				
480	46.5 × 22.5 × 31.5	525	508	na
Surefire				
101B	45 × 26.3 × 53	1000	440	160,000
Summeraire				
S–110G	46 × 24.3 × 32.5	1200	420	120,000
Thermo-Control				
200A	39 × 25 × 37	na	260	70,000
400A	39 × 31 × 38	na	380	95,000
500A	45 × 31 × 44	na	480	135,000
Thermo Pride				
W/C 20	43.3 × 25 × 50.3	700/1000	635	90,000
W/C 27	46.8 × 27 × 58.5	1100/2200	850	130,000
Twentieth Century				
1251	68 × 44.6 dia.	na	1243	110,000
Timber Iron	32 × 23 × 50	na	600	100,000
Valley Forge				
401	47 × 27 × 27	920	550	100,000
Woodchuck				
2300	38.5 × 24 × 28	500	445	110,000
2600	42 × 26 × 28	930	495	140,000
2800	42 × 26 × 28	930	505	120,000
3100	48 × 26 × 31	930	555	160,000
Yukon				
Klondike	50.8 × 30 × 62	800/1400	720	131,000

ASHLEY. *(wood)* U.S. A steel heat exchanger and firebox with cast iron liners and thermostatically controlled down-draft system. Ash pan, large feed door, blower, air filter, and cabinet. Duct size 15.5″ × 14″.

CHARMASTER. *(wood)* U.S. Fabricated of a 5/16″ steel plate firebox and 14 gauge steel heat exchanger. 11″ × 13″ fuel door, thermostatic draft control, ash door, and jacket. 24″ × 28″ warm air plenum and 16″ × 20″ cold air return.

COUNTRY COMFORT. *(wood/coal)* U.S. Designed using 10 gauge steel firebox with firebrick liner, grates (wood or coal) 13″ × 10″ fuel door, oversize ash pan, thermostatically controlled damper, and air distribution blower. 14.5″ × 19.3″ duct size with optional filter box and cold air return.

CROSSWINDS. *(wood/coal)* U.S. Manufactured with a 10 gauge heat exchanger, firebrick-lined firebox and airtight door. Motorized damper, steel grate, ash pit, electric controls, blower box with filter and jacket.

DANIELS. RCD Series — *(coal)* U.S. Constructed of 10 gauge electrically welded steel firebox and horizontal drum heat exchangers. 10.5″ × 14.5″ feed door, shaker grate, 3000° F. firebrick liner, optional damper motor and controls and blower with housing if forced air circulation is desired. R W Series — *(wood)* U.S. Built with 12 gauge steel firebox and heat exchanger drums. 15″ × 17″ feed door, ash pit, grates, automatic controls, and sheet metal jacket. Units available as gravity, pipeless, or blower models.

DORNBACK. *(wood/coal)* U.S. 12 gauge vertical heat exchanger with firebrick lined firebox and cast iron grates. Large feed door, ash pit, ¼ H.P., 9″ × 9″ blower, air filter and automatic draft control. 22″ × 23″ warm air plenum opening and 13″ × 23″ cold air return opening.

DUO-MATIC. *(wood/coal)* U.S. Welded steel primary and secondary heat exchangers over firbrick-lined firebox. Shaker grates, ash pit, 16.8″ × 13.5″ fuel door and insulated jacket. Optional blower and filter section. Hot air plenum, 24.1″ × 31.9″, and cold air plenum, 19.3″ × 18.1″.

ENERGY KING. *(wood)* U.S. A 3/16″ steel plate furnace with secondary heat exchanger and insulated outer shell. Sliding secondary heat baffle, automatic thermostat, 12″ × 12″ feed door, firebrick liner, cast iron grates, and blower filter box. 10″ diameter hot air outlet. 5-year warranty.

HOT SHOT. *(wood)* U.S. 10 gauge steel firebox and heat exchanger with firebrick liner and cast iron grate. 12″ × 13″ feed door, ash pan, manual draft, baffle and blower box with air filter. 16″ × 18″ hot air plenum opening.

JOHNSON ENERGY SYSTEMS. *(wood)* U.S. Uses a baffled 7 gauge steel firebox with a 12″ square feed door. Model 8800 has a grate and ash drawer. Thermostatically controlled damper, smoke deflector and 16 gauge jacket. Two- 7″ diameter heat outlets, no cool air return plenum but contains an air filter.

KERR CONTROLS. *(wood)* Canada. Uses 1/8″ steel plate primary and 14 gauge rolled steel secondary combustion chambers with preheated draft air. Large fuel door, damper motor primary draft, air filter and jacket. 5-year warranty. Large warm air and cold air plenums.

KEYSTONER. *(wood/coal)* U.S. 3/16″ steel firebrick-lined firebox and 11 gauge heat exchanger. Models 100 and 120 have a stainless steel grate, Model 170 has a cast iron shaker grate. 13″ × 15″ feed door, ash pan and jacket. Large supply plenum, no return plenum or air filter.

KICKAPOO. *(wood)* U.S. Refractory lined steel firebox with large cast iron feed door.

Ash pan, manual draft control, jacket, and optional blower housing with filter. 19″ × 17″ output plenum.

Kodiak. *(coal/wood)* U.S. A firebrick-lined steel firebox with large feed door, shaker grates, and ash pit. Motorized primary damper, safety controls, air filter, and jacket. Large supply and return plenum openings.

Newmac. *(wood/coal)* U.S. All welded steel construction with firebrick-lined firebox, shaker grates, and ash pit for coal. Preheated combustion air introduced by thermostatically controlled blower, 14″ square fuel door, air filter, jacket, and safety controls. Large air supply and return air duct connections.

Oneida royal. *(coal)* U.S. Firebrick-lined firebox, steel heat exchanger in round or rectangular jacket. Shaker grate, large fuel and ash doors, manual or thermostatically controlled draft, and large firebox. Space for several air outlet ducts.

Passat. *(wood)* Manufactured in Denmark. A large, heavy steel cylindrical firebox that will burn three-foot fuelwood. Large and small fuel doors, bimetallic thermostat for draft control, baffled blower and housing. 15.8″ square exhaust plenum, 17.7″ square intake plenum. Available with oil burner backup or stoker.

Pribbs. *(wood)* U.S. A 3/16″ steel plate, double drum furnace that will burn four-foot cordwood. Manual draft, blower, and limit switch. Baffled firebox and secondary heat exchanger. Heat can be ducted.

Riteway. *(wood)* U.S. Heavy steel heat exchanger above a fireclay-lined firebox. Ash pan, grate, draft blower, air filter, and galvanized jacket. Air is preheated before passing through heat exchanger. Includes mounting tube for oil/gas burner.

Russell. *(wood/coal)* U.S. Features a baffled firebox with preheated secondary air. Firebrick-lined, shaker grate, ash pan, 3500-square-inch heat exchanger, 11.5″ square cast iron feed door, manual draft controls, and jacket. Large heat outlet connection, no return air connection or filter.

Seville. *(wood/coal)* U.S. Large firebrick-lined firebox with shaker grates and interior baffled for secondary combustion. Thermostatically controlled combustion air fan, ash pan, and jacket. 8″ diameter air outlet, filter or connection to blower. 5-year warranty.

Shenandoah. *(wood/coal)* U.S. A heavy gauge aluminized steel firebox with firebrick lining and cast iron shaker grate. Smoke baffle, door-mounted thermostatically controlled damper, manual blower switch for summer cooling, ash pan, and jacket. 20″ square warm air plenum and 15″ × 20″ cold air plenum and filter. Available without blower for add-on installation.

Spartan. *(wood/coal)* U.S. Features a 1/4″ steel plate firebox lined with firebrick and a 10 gauge welded steel heat exchanger. Large firebox, cast iron shaker grate,

60-cfm thermostatically controlled draft fan, smoke bypass, large ash pan, and automatic controls. 16″ square supply plenum, no filter or cold air return plenum.

Steel king. *(wood/coal)* U.S. Similiar in design and construction to the Spartan.

Surefire. *(wood)* U.S. Uses an 8 gauge steel firebox, 11 gauge labyrinth-style heat exchanger, and secondary combustion chamber. Equipped with damper motor, controls, 14″ × 12″ fire door, and jacket. Hot air plenum, 31″ × 24″; cold air plenum, 18″ × 24″, and filter.

Summeraire. *(wood/coal)* Canada. Features labyrinth heat exchanger, 8 gauge firebox and optional cast iron shaker grates and firebox liner. Ash cleanout door, large feed door, damper motor, control package, and jacket. Warm air plenum, 22″ square; return air plenum, 22″ square, and filter.

Thermo control. *(wood)* U.S. Heat exchanger welded onto firebrick-lined firebox. Large airtight fuel door, preheated primary and secondary air, easy-to-assemble plenum kit, optional electric damper, and cold air return connection. No blower.

Thermo pride. *(wood/coal)* U.S. Contains a firebrick-lined firebox under a 10 gauge copper-coated steel heat exchanger. Cast iron shaker grates, ash pit, 10″ high fuel door, secondary heat exchanger, motorized damper, controls, and jacket. Large warm air and cold air plenums, belt-driven blower, and filter. 10-year warranty.

Twentieth century. Model 1251 — *(coal/wood)* U.S. A large, round, cast iron furnace that will operate by either gravity or forced-air circulation. Round cast iron firebox, shaker grate, cast iron heat exchanger, large feed and ash doors, and sheet metal jacket. Features blast furnace primary draft.

Timber iron. *(coal/wood)* U.S. Cast iron firebox and heat exchanger. Large ash pan, dump grates, 13.5″ × 9″ fuel door, motorized thermostatically controlled draft damper, secondary air, belt-driven blower, filter, and large plenum.

Valley forge. *(coal/wood)* U.S. Has a firebrick-lined firebox and steel heat exchanger. Large ash pan, cast iron shaker grates, optional automatic draft control, dual blowers, filter box, large heat plenum, and insulated jacket. 5-year warranty.

Woodchuck. *(wood/coal)* U.S. A heavy steel and firebrick-lined unit with secondary heat chamber on models 2800 and 3100. Ash pan, large fuel door, and jacket. Optional shaker grate for coal, draft blower and filter box. 23″ × 20″ warm air plenum.

Yukon. *(wood/coal)* U.S. Extra large heat exchanger above a firebrick-lined firebox. Secondary air, large fuel door, optional shaker grate, belt-driven blower, filter, and insulated cabinet. 20″ × 24″ heat plenum and 16″ × 20″ cold air return plenum.

ADD-ON FURNACES

The following is a listing of a variety of solid-fuel furnaces that are designated by the manufacturer as primarily for add-on installations.

Manufacturer and Model	Dimensions Height by width by depth	Blower Capacity Cu.ft./Min.	Weight pounds	Heating Capacity Btu/hour
Birmingham				
724 U	35.3 × 21 × 30.5	520	265	50,000
Brunco				
110	49 × 24 × 31	525/1050	550	na
160	49 × 24 × 40	1500	685	na
Daka				
451C	37 × 22 × 24	465	325	62,000
501	44 × 24 × 27	930	400	96,000
601	48 × 24 × 32	930	425	114,000
Defiance				
Volcano II	42.5 × 23.8 × 41.6	465	585	120,000
Dover				
WB 26	46 × 26 × 30	na	700	100,000
WB 30	46 × 26 × 30	na	700	130,000
Hunter				
Mini Furnace	48 × 21 × 55	1200	300	85,000
Jensen				
24A	40 × 23 × 24	500	330	na
30A	40 × 23 × 30	500	375	na
Kickapoo				
ADD 2	42.3 × 24.5 × 31.4	525	338	na
Leaders				
LH 25	42 × 26 × 29	500	350	na
LH 30	46 × 24 × 30.5	1000	455	na
Longwood				
MK II	40 × 24 × 54	750	450	65,000
Monarch				
Wood AF 324A	42 × 22 × 32	500	300	48,000
Wood/coal				
AF 424A	42 × 22 × 47.4	900/1385	400	48,000
Wood/coal				
AF 524A	48 × 22 × 32	1385	366	72,000
Wood				
AF 524AX	48 × 22 × 32	1385	375	72,000
Surefire				
201A	45 × 26.5 × 32.3	na	390	140,000
Summeraire				
S-100G	46 × 24.3 × 32.5	na	335	120,000
Woodmaster				
	36.2 × 22 × 30.3	555	225	na

Birmingham. *(wood)* U.S This model features a cast iron firebox and shaker grate, steel heat exchanger, ash pan, 13″ × 10″ fuel door, and motorized draft damper. Outlet hot air duct diameter, 8″, return air diameter, 6″.

Brunco. *(wood/coal)* U.S. Furnace is airtight, steel plate design with shaker grates and firebrick liner. Large ash pan, fuel door, secondary heat exchanger, draft fan, safety controls, and jacket. Outlet duct, 12″ diameter. No return duct connection.

Daka. *(wood/soft coal)* U.S. Furnace features heavy 7 gauge steel combustion chamber with firebrick liner. Bimetallic automatic draft control on Models 451C and 501, combustion air blower on Model 601, ash pan, cast iron grate, full width baffle, and large fuel doors. 10″ diameter outlet flue and optional cold air filter box. Soft coal shaker grate optional.

Defiance. *(wood/coal)* U.S. A steel plate, double-pass baffled furnace with large firebox and fuel door. Ash pit, preheated combustion air with three-position manual draft control. Blower takes room air and exhausts it through an 8″ × 14″ outlet.

Dover. *(wood)* U.S. A large tube type, heavy duty steel plate furnace with refractory liner. Contains a baffle, ash pit, grate, blower with filter, large loading door, draft control, and jacket.

Hunter. *(wood)* Canada. Built with a stainless steel firebox, baffle, airtight door, thermostatically controlled motorized damper, air filter, prewired controls, and insulated jacket.

Jensen. *(wood/coal)* U.S. A steel plate furnace with secondary combustion chamber and cast iron heat baffles. Large feed door, ash pan, shaker grate, firebrick combustion chamber, jacket, 37-cfm draft fan, and safety controls. No return air connection.

Kickapoo. *(wood)* U.S. Heavy steel firebox lined with firebrick. Cast iron feed door with manual draft control, baffle, blower, and jacket. Heat outlet, 8″ diameter, return air duct can be attached to blower. No filter.

Leaders. *(wood)* U.S. Firebrick-lined steel firebox and firebrick grate. Baffled firebox with thermostatically controlled draft blower to provide primary and secondary air. Safety controls and jacket. 12″ diameter heat outlet and non-filtered air blowers.

Longwood. *(wood)* U.S. Features a 32″ long cylindrical firebox made from 3/16″ steel with cast iron liners. Large feed and ash doors, motorized, thermostatically controlled combustion air damper, bypass damper, air filter, and jacket. Supply and return air duct outlets, 18″ × 8″.

Monarch. *(wood/coal)* U.S. Has large cast iron firebox with grates and heavy duty heat exchanger. Includes ash pan, thermostatically controlled damper, flue deflector, jacket, controls, and large fuel door. Warm air plenum, 12″ × 16″ (14″ × 20″

on AF 524A), cold air return duct connection, and filter. Model AF 524AX contains a catalytic combustion system for increasing efficiency.

SUREFIRE. *(wood)* U.S. Same as Surefire 101B but without blower.

SUMMERAIRE. *(wood/coal)* Canada. Similar to Summeraire Model S110G furnace except without blower box and filter.

WOODMASTER. *(wood)* U.S. Utilizes a 14 gauge heat exchanger and firebrick-lined firebox. 10" × 12" fuel door, ash door, bimetallic thermostatically controlled damper, smoke curtain, and jacket. Optional blower.

MULTIFUEL FURNACES

The following is a listing of some of the multifuel furnaces. These units burn either wood or coal as the solid fuel and use fuel oil, natural, or LP gas or electricity as the conventional fuel.

Manufacturer and Model	Dimension Height by width by depth	Blower Capacity Cu.ft./Min.	Weight pounds	Heating Capacity Btu/hour
Charmaster				
Wood/oil	54 × 28 × 55	1780	680	170,000
Wood/gas	54 × 28 × 55	1780	849	125,000
Combo				
	53.5 × 28.5 × 57.3	1025	585	126,000
Dornback				
CWOG 70-0	48 × 54 × 25	na	470	100,000
CWOG 70-G	48 × 54 × 25	na	440	100,000
Dual-Aire				
DA150	54.5 × 28 × 75	1725	900	150,000
Duo Matic				
BBC	47.8 × 51.3 × 57.5	1016	760	95,000
CWO-B 112	51.3 × 44.1 × 48.3	1200	970	112,000
CWO-B 140	51.3 × 44.1 × 48.3	1440	970	134,000
Hunter				
	53 × 48 × 56	na	800	165,000
Longwood				
	40 × 50 × 69	1470	na	na
Newmac				
Wood CL85/95	42 × 50 × 46	900/1000	700	82,000/ 94,000
Wood CL115C/ 170C	51.5 × 48.3 × 53.5	1300/1900	1035	111,000/ 169,000

Manufacturer and Model	Dimension Height by width by depth	Blower Capacity Cu.ft./Min.	Weight pounds	Heating Capacity Btu/hour
Wood/coal				
CL115G/170G	51.5 × 48.3 × 53.5	1300/1900	1000	111,000/ 169,000
Oneida Royal				
Wood AWGO				
80/125	47.3 × 27.8 × 62	na	na	80,000/ 125,000
Coal ACO				
80/112	59.5 × 52.8 × 42.5	na	650/775	85,000/ 112,000
Riteway				
LF 20	53 × 34 × 68	na	1000	125,000
Surefire				
Oil 101 CO	53 × 42.5 × 45	1200	660	140,000
Electric 101 CE	54 × 26.3 × 53	1200	480	160,000
Thermo Pride				
W/C				
20/OL5-85	43.3 × 50 × 50.3	800/1200	635/410	90,000
W/C				
27/OL11-112	46.8 × 52 × 58.5	1100/2200	850/500	130,000
W/C				
27/OL16-125	46.8 × 54 × 58.5	1100/2200	850/560	130,000
Yukon				
Husky LWO				
85/112	61.9 × 30 × 65	800/1400	869	85,000/ 112,000
Polar LWO				
151/168	50.8 × 30 × 58.5	1200/1800	1072	151,000/ 168,000
Wood/electric	50.8 × 30 × 52	800/1400	875	68,260

CHARMASTER. *(wood/oil/gas)* U.S. Similar to solid fuel furnace but with conventional burner attached.

COMBO. *(wood/oil)* U.S. Large firebox with the oil burner above the fuel door. Automatic damper control, ash pan, controls, belt-driven blower, jacket and large warm air plenum.

DORNBACK. *(wood/coal/oil/gas)* U.S. Features separate fireboxes and 12 gauge steel heat exchangers. Large feed door and ashpit doors, belt-driven blower, filter, firebrick liner, shaker grate, and electronic draft control. 23″ × 22″ warm air discharge opening and 23″ × 13″ return air opening. High efficiency burners.

DUAL-AIRE. *(wood/coal/oil)* Canada. Horizontal firetube heat exchanger made from 12 gauge steel. Firebrick-lined firebox, cast iron shaker grates, separate oil burner combustion chamber, 14″ square fuel door, ash pit, adjustable speed blower, filter, and electric damper motor. Hot air plenum, 22.3" × 35.3" and cold air plenum, 18.3" × 22.3".

DUO-MATIC. *(wood/oil)* U.S. A base burner with fiber ceramic chamber for the oil burner. Steel heat exchanger, firebox sides and bottom lined with firebrick, large dual-purpose fire and ash pit door, blower section with filter, and motorized damper air intake. Hot air plenum, 24.3", cold air plenum 23.3" square. Oil burner can be increased to 136,000 Btu/hour output.

DUO-MATIC SERIES CWO. *(wood/coal/gas/oil)* U.S. Separate combustion chamber furnace of heavy duty welded construction. Blower section has twin blowers, firebrick-lined firebox, shaker grates, ash pit, damper motor, and safety control. Hot air plenum, 24" × 46.1", cold air plenum, 16.1" × 46.1".

HUNTER. *(wood/oil)* Canada. Separate baffled combustion chambers and common heat exchanger. Wood and oil can be operated independantly or simultaneously. 16" × 13" feed door, twin thermostats, twin blowers, stainless steel wood firebox, filter, and safety controls. 26" × 54" warm air plenum, 17" × 54" cold air plenum.

LONGWOOD. *(wood/oil/gas)* U.S. utilizes a 14 gauge stainless steel combustion chamber liner within a 3/16" boiler plate steel cylindrical heat exchanger. Takes five-foot logs. Ignition and burning rate controlled by oil/gas burner. Draft air provided by burner. 12" square fire door, heat reclaimer, and aluminum jacket. Large supply and cold air return outlets with filter.

NEWMAC. *(wood/coal/oil)* U.S. Separate combustion chambers with independently controlled firing. Steel heat exchanger, firebrick/cast iron firebox, grates, and ash pit on coal unit, jacket thermostatically controlled combustion air blower, filter, and large heat duct connections.

ONEIDA ROYAL. *(wood/coal/oil/gas)* U.S. These series of furnaces with different size oil/gas burners have a heavy steel heat exchanger above a grated firebox. The solid fuel is ignited by the oil/gas burner. Air movement provided by blower with filter. Automatic draft regulation and safety equipment are standard. 10-year warranty.

RITEWAY. *(wood/coal/oil/gas)* U.S. Includes cast iron gas combustion flue, heat exchanger, draft inducer, draft blower, firebrick liner, ash pan, shaker grate, hot air blower with filter, and safety controls. 16" × 30" warm air plenum.

SUREFIRE. *(wood/oil/electric)* U.S. Furnace unit same as Surefire Model 101B but with oil burner, 10 kilowatt or 20 kilowatt electric heater. 10-year warranty.

THERMO PRIDE. *(wood/coal/oil/gas)* U.S. These units are a combination of the solid fuel unit described earlier and a companion oil or gas furnace with controls and adapter for the warm and cold air outlets.

YUKON. *(wood/coal/oil/gas/electric)* U.S. Unit similar to Klondike model with the addition of separate ceramic firebox for oil/gas. Several sizes of conventional burners available. Automatic switching from solid fuel to conventional fuel.

SOLID FUEL BOILERS

The following is a listing of some of the styles and types of solid fuel boilers. Both steam and hot water units are available. Most of these can be installed as add-on units.

Manufacturer and Model	Dimensions Height by width by depth	Boiler Capacity Gallons	Weight pounds	Heating Capacity Btu/hour
Buderus				
Logana				
2.40 — 14	38.8 × 19.3 × 15.1	6.3	450	50,000/
19	38.8 × 19.3 × 19.1	7.4	532	68,000/
				76,000
27	38.8 × 19.3 × 23	8.6	620	96,000/
				108,000
35	38.8 × 19.3 × 27	10	706	125,000/
				140,000
40	38.8 × 19.3 × 12.1	12.1	876	145,000/
				160,000
Columbia				
L-22-C	53 × 26 × 27	50	1050	139,000
Dedietrich				
CF 124CS	41.1 × 19.8 × 15.9	8.2	561	78,000
CF 125CS	41.1 × 19.8 × 20.2	9.8	667	95,000
CF 126CS	41.1 × 19.8 × 24.6	11.4	772	110,000
CF 127CS	41.1 × 19.8 × 28.9	12.9	878	125,000
CF 128CS	41.1 × 19.8 × 33.2	14.5	983	145,000
CF 129CS	41.1 × 19.8 × 37.6	16.1	1089	160,000
Energy King				
	47 × 22 × 33	26	750	90,000
Energy Mate				
7836-B	36 × 22 × 26	21	500	na
7848-B	48 × 24 × 30	27	650	na
Eshland				
C40	44 × 28 × 36	40	1050	120,000/
				150,000
C55	54 × 28 × 36	55	1220	180,000/
				225,000
Firecraft				
CW50	50 × 23.5 × 23	30	645	140,000
FWB 140	45.5 × 27 × 32	35	875	140,000
Franco Belge				
91-315W	50.5 × 20.5 × 17	na	630	96,000
91-415W	50.5 × 20.5 × 21	na	749	130,000
91-515W	50.5 × 20.5 × 25	na	864	162,000
91-615W	50.5 × 20.5 × 29	na	981	194,000

Manufacturer and Model	Dimensions Height by width by depth	Boiler Capacity Gallons	Weight pounds	Heating Capacity Btu/hour
Frank's Piping				
W-2	48 × 24 × 40	55	895	123,000
W-3	56 × 27 × 40	65	930	200,000
Froling				
FH 25	33.5 × 19.3 × 28.7	18.8	616	99,000
FH 30	39.8 × 19.3 × 32.7	23.3	792	127,000
FH 40	42.1 × 21.3 × 32.7	28.3	990	159,000
FH 50	50.4 × 23.2 × 33.7	33.8	1144	218,000
F 20	48 × 22.3 × 25.4	na	473	54,000/ 68,000
F 25	48 × 22.3 × 25.4	na	484	68,000/ 85,000
F 36	48 × 22.3 × 25.4	na	484	95,000/ 123,000
F 45	48 × 22.3 × 28.3	na	550	123,000/
Gyro				
FF 25	37.2 × 33.1 × 49.2	26.4	517	100,000
FF 30	37.2 × 33.1 × 52.5	33	693	138,000
Kerr				154,000
Jetstream				
120 SB	50 × 24 × 42	50	1490	120,000
Kerr				
Titan — wood	43.5 × 26.5 × 35.5	na	675	140,000
Titan — coal	58 × 28 × 40	35	825	100,000
Kinetic Solar Burners				
KB 80 wood	47 × 26 × 30	21	620	120,000
KB 80C coal/wood	47 × 26 × 30	21	793	160,000
Kopo				
KB 18	40.9 × 27.2 × 20.5	18.5	595	61,000
KB 25	42.1 × 37.8 × 26.8	26.4	968	92,000
KB 45	41.3 × 37.8 × 38.2	37	1210	154,000
Mascot				
MW 100 wood	28 × 24 × 28	12	570	85,000
MC 200 coal	45 × 24 × 28	12	900	110,000
Memco				
MW 100	48 × 22 × 32	50	650	140,000
New England Boiler				
WC 120	50.5 × 25 × 41.5	na	1301	120,000
WC 150	50.5 × 25 × 49.5	na	1652	165,000
Newmac				
B 160 wood	48 × 24 × 30.5	65	800	160,000
BC 160 wood/coal	59 × 24 × 30.5	65	970	160,000
Northland Boiler				
W 620	47.3 × 21 × 27.1	56	986	110,000
W 720	47.3 × 21 × 34.5	66	1130	147,000

Manufacturer and Model	Dimensions Height by width by depth	Boiler Capacity Gallons	Weight pounds	Heating Capacity Btu/hour
Oneida Royal				
CWB 112	49.5 × 21 × 27.1	60	986	112,000
CWB 150	49.5 × 21 × 34.6	70	1130	150,000
Passat				
HO-20	35.4 × 29.5 × 47.6	27.7	518	72,000
HO-30	35.4 × 29.5 × 58.2	34.3	629	108,000
HO-35	35.4 × 29.5 × 68.1	42	705	132,000
HO-45	47.3 × 39.4 × 67.3	46.2	1215	152,000
Prairie Wood				
	42.5 × 20 × 24	33	na	na
Quaker				
10-15-3	50.5 × 20.5 × 16.9	na	na	96,000
10-15-4	50.5 × 20.5 × 20.9	na	na	130,000
10-15-5	50.5 × 20.5 × 24.9	na	na	162,000
Riteway				
LB 30	66 × 36 × 56	na	2000	160,000
LB 50	66 × 36 × 68	na	2600	200,000
Simplex Multitherm				
HF 120	52.1 × 22.5 × 25.5	na	1040	120,000
Stadler				
H 33	41.6 × 24.8 × 31.1	25	726	132,000
H 45	46 × 24.8 × 33.3	29	836	180,000
H 60	51.7 × 24.8 × 35.6	40	1078	240,000
Tasso				
U 3	45.8 × 26.3 × 26	19	718	68,000
Tritschler				
HK 2516	42.5 × 17.1 × 18.8	9	395	62,000
HK 3020	42.5 × 17.1 × 22.7	12.5	425	78,000
HK 3825	42.5 × 19.2 × 22.7	15.8	483	98,000
HK 4832	42.5 × 20.8 × 25	24.5	558	125,000
Woodbine				
1500	36 × 27 × 36	16.5	na	140,000

BUDERUS. *(wood/coal)* West Germany. A cast iron sectional boiler suitable for solid-fuel firing but adaptable to oil/gas. Wet base design. Flue box with damper and damper positioner, loading door, ash door with damper, deep ash pan, threaded flanges for water flow, cleaning brushes, aquastat, and insulating jacket.

COLUMBIA. *(coal/wood)* U.S. A water tube, water leg hot water, or steam boiler. Large fire and ash pit doors, manual draft control, shaker grates, and insulated jacket.

DEDIETRICH. *(wood/anthracite/coke)* West Germany. Sectional cast iron boiler convertible to oil/gas. Fuel and ash doors with air dampers, aquastat, shaker grates, ash pan, cleaning tools, assembly and operation manual. Features wide fuel door and inner flue ways designed to limit soot and creosote accumulation.

ENERGY KING. *(wood)* U.S. A firebrick-lined, steel-jacketed boiler featuring simple construction and design. Contains optional shaker grates, ash pan, manual and automatic draft control, heat baffle, and insulated outer shell.

ENERGY-MATE. *(coal/wood)* U.S. 1/4″ steel plate, water wall boiler with firebrick-lined firebox and cast iron grates. Thermostatically controlled draft blower, ash pan, large fuel door, bypass damper, safety controls, and jacket. Adapts easily to Energy-mate stoker.

ESHLAND. *(coal/wood)* U.S. A firebrick-lined, steel boiler featuring two automatic draft systems, one for each fuel. Contains ash pan, removable baffle, safety controls, and large loading door. The triangular rotary grate system allows adjustment of the air depending on the fuel used.

FIRECRAFT. *(coal)* — CW50 U.S. A wet leg boiler with airtight construction and heavy duty shaker grate. Automatic, thermostatically controlled dampers, secondary air, ash pit, and safety controls. The FWB-140 is wood fired and has similiar design features.

FRANCO BELGE. *(wood/coal)* U.S. Cast iron, vertical section, wet base construction. Thermostatically controlled primary air damper, shaker grates, ash pan, large fuel door, and insulated jacket. Shipped assembled. All models can be purchased for steam operation.

FRANK'S PIPING. (wood/coal) Canada. Large steel plate combustion chamber will hold 90 pounds (W-2) or 160 pounds (W-3) of dry wood for long burn cycle. Motorized air damper, ash pan, insulated steel jacket and cleaning tool. Shaker grates available for coal burning. Similiar model 15 psi steam boilers are available.

FROLING. FH series — *(wood/coal/coke)* West Germany. Steel vertical tube boiler operates with downward burning principle. Water-cooled grates, magazine fuel chamber, top feed door, bimetallic aquastat-damper control, and insulated jacket. Can be operated with low draft chimneys.
F series — *(coal/coke/wood)* West Germany. Wet leg and water-cooled grate designed boiler. Large feed doors with damper, secondary heat exchanger, insulated jacket, and adjustable supports.

GYRO. *(wood)* Denmark. The heavy steel horizontal cylindrical firebox takes three-foot pieces. Large fire door with intake dampers thermostatically controlled and insulated fire jacket.

KERR. *(wood)* Canada. A high efficiency boiler utilizing a firetube heat exchanger with turbulator inserts. Includes refractory firebox, vertical fuel tube, draft inducer, and preheated secondary air. Designed around the Hill concept, it is generally connected to a large hot water storage tank.

KERR. *(wood/coal)* Canada. A dry base boiler with firebrick lined combustion chamber and vertical firetubes. Thermostatically controlled damper, preheated secondary air, shaker grate, and unique steam blow-down safety system.

KINETIC SOLAR BURNERS. *(coal/wood)* U.S. A wet leg boiler with firebrick and cast iron firebox. Features dump grates, automatic draft control, preheated secondary air, ash door, and smoke curtain. Pyroseal door gaskets provided long life, airtight design.

KOPO. *(chips/pellets/wood)* U.S. A hopper feed boiler with refractory flame chamber. Combustion air blower provides both primary and secondary air. Feed can be automated from storage bin.

MASCOT. *(wood/coal)* U.S. A wet leg, firebrick-lined firebox with adjustable baffle. Damper motor controlled draft, large feed door, safety controls, and optional jacket. The coal unit has shaker grate and ash pit set below the firebox.

MEMCO. *(wood/coal)* U.S. Utilizes 1/4″ steel plate construction, wet leg design and double shaker grate system. Ash pan, adjustable legs, large fill door, and electric control system.

NEW ENGLAND BOILER. *(wood/coal)* U.S. A horizontal tube, steel boiler with wet leg water jacket. Cast iron shaker grates, insulated jacket, expansion tank, automatic draft control, barometric draft control, and cleaning brush. 5-year limited guarantee.

NEWMAC. *(wood/coal)* U.S. Horizontal tube boiler with water backed firebox walls. Large loading doors, blower primary draft, safety controls, prewired electrical central control panel and insulated jacket. Model BC-160 is for wood/coal operation.

NORTHLAND BOILER. *(wood/coal)* U.S. Horizontal firetube, wet leg design boiler with large water volume. Airtight fuel and ash doors, cast iron grates, ash pan, expansion tank, safety controls, and cleaning brushes. 5-year warranty. Available as steam units.

ONEIDA ROYAL. *(coal/wood)* U.S. Heavy steel plate, wet leg boiler with horizontal firetube construction. Large feed door, rocker-type shaker grates, insulated jacket, automatic fire draft regulator, barometric draft control, and cleaning brushes.

PASSAT. *(wood/straw)* Denmark. Heavy steel cylindrical firebox surrounded by water jacket. Firebox will take three-to four-foot firewood, double-draft doors, non-electrical thermostatic control, rear heat exchanger and baffles. Can be fitted with stoker or integral oil burner.

PRAIRIE WOOD. *(wood/coal/electric)* U.S. A 3/16″ steel plate, water wall boiler with horizontal firetubes. Cast iron grates, damper motor, safety controls, ash pit, large loading door, and insulating jacket. Optional electric heating element.

QUAKER. *(coal/coke/wood)* U.S. A wet base, cast iron boiler with deep firebox and rocking type shaker grates. Ash pan, insulated jacket, and optional damper regulator and barometric draft control.

RITEWAY. *(wood/coal)* U.S. A heavy steel water wall boiler with primary air blower and secondary air jets. Large fuel door, grates, ash pit, fuel selector damper, draft inducer, safety controls and jacket. Has a connection for oil or gas burner. Steam boilers also available.

SIMPLEX MULTITHERM. *(wood/coal)* U.S. A three-pass, pitched horizontal watertube boiler with wet leg construction. Shaker grates, bimetallic aquastat damper control, barometric draft control, and insulated jacket. Also available for steam operation.

STADLER. *(wood/coal)* West Germany. A wet base, vertical tube boiler with magazine feed. Top feed door, ash pit, insulated jacket, primary and secondary air supply. Equipped with safety coil for overheat protection.

TASSO. *(wood/coal)* Denmark. A cast iron sectional boiler that utilizes a downdraft gasification combustion principle. Fixed or shaker grates, bimetallic aquastat-damper control, controls and jacket. Available in several sizes to 300,000 Btu/hour.

TRITSCHLER. *(coal/wood)* West Germany. A wet wall steel boiler with turbulator baffle and bar grate. Large firebox and feed door, ash pit and foam insulated jacket. Features safety heat sink which absorbs overheat above 212° F.

WOODBINE. *(wood/coal)* U.S. Steel plate, wet wall boiler with shaker grates and large firebox. Slant air-cooled feed door, ash pan, damper motor, and adjustable baffle.

ADD-ON BOILERS

The following is a listing of some of the boilers that have been designated by the manufacturer as primarily for add-on installations. These units burn wood or coal as the solid fuel and oil, gas or electricity as the conventional fuel.

Manufacturer and Model	Dimension Height by width by depth	Water Capacity Gallons	Weight pounds	Heating Capacity Btu/hour
Duo-Matic				
WBA	63.5 × 26.8 × 32.5	na	900	120,000
Energen				
101	42.3 × 21.5 × 29	na	385	102,000
Energy-Mate				
7836B	36 × 22 × 26	21	500	na
7848B	48 × 24 × 30	27	650	na

Manufacturer and Model	Dimension Height by width by depth	Water Capacity Gallons	Weight pounds	Heating Capacity Btu/hour
Furnace Works				
SFB 3	27 in. dia × 30	9.7	235	140,000
Hoval				
HK 30	45 × 26 × 28	na	na	80,000/ 120,000
HK 45	55 × 26 × 36	na	na	128,000/ 180,000
Hydra Fire				
	48 × 22 × 34	2	525	140,000
Jensen				
24A	40 × 23 × 24	na	580	na
30A	40 × 23 × 30	na	625	na
Menominee				
MBW 8000	46 × 24 × 31.5	26	760	125,000
Monarch				
wood AF-B101	50 × 19 × 32	4.7	670	76,000
coal AF-B101C	50 × 19 × 32	4.7	670	58,000
NECA				
LC-18	40 × 20 × 16.8	7.2	501	72,000
LC-27	40 × 20 × 20.5	8.7	587	108,000
LC-32	40 × 20 × 24.2	10.2	673	128,000
LC-38	40 × 20 × 27.9	11.7	759	152,000
Oneida Royal				
ACWB 100	41 × 24.8 × 29.5	na	720	na
Russell				
	42 × 22.5 × 28	na	800	70,000
Severance				
	38 × 24 × 24	na	480	110,000
Seville				
AB 80	40 × 23 × 24	14	550	80,000
AB 120	46 × 23 × 24	17	650	120,000
AB 160	46 × 23 × 36	20	730	160,000
Tarm				
202	37 × 18 × 19.8	24	450	80,000
303	51 × 23 × 23	21	645	120,000
Thermo-control				
200SC	27 × 18 × 33	na	245	70,000
400SC	27 × 24 × 34	na	360	95,000
500SC	33 × 24 × 40	na	460	135,000
Timber Eze				
210	42 × 25 × 30	12	775	90,000
220	48 × 25 × 36	19	875	120,000
UMCO				
WC 90	41 × 24.5 × 25.5	17.9	640	90,000
WC 120	45 × 24.5 × 25.5	22.5	680	120,000
Valley Forge				
701	50 × 27 × 33	50	950	140,000

DUO-MATIC. *(wood)* U.S. A firetube steel plate boiler with firebrick-lined firebox. Large cast iron doors with damper motor draft control, ash pan, insulated jacket, aquastat, and pressure gauge.

ENERGEN. *(wood/coal)* U.S. A top-loading, dry base boiler with steel pipe heat exchanger. Refractory firebox, ash pit, damper motor draft control, shaker grate, and aquastat. Model 102 can be added to a hot-air furnace with a heat exchanger placed in the plenum.

ENERGY-MATE. *(wood/coal)* U.S. 1/4" steel plate, wet wall boiler with firebrick-lined firebox and cast iron grate. Thermostatically controlled draft blower, ash pan, controls, insulated jacket, and 5-year warranty. Can be adapted to Energy-mate stoker for coal.

FURNACE WORKS. *(wood)* U.S. Features two concentric steel cylinders with fire inside and water between. 13.5" square fuel door, thermostatically controlled damper motor, safety controls, and 5-year warranty.

HOVAL. *(wood/coal/coke)* Liechtenstein. Magazine feed, top-loading combustion chamber has refractory-lined firebox. Vertical firetubes, shaker grate, insulated jacket, and large cleaning door. Stovepipe collar on back near base.

HYDRA FIRE. (coal/wood) U.S. Incorporates a 13-square-foot tube plate heat exchanger above a refractory-lined combustion chamber. Motorized damper primary air, secondary air ducts, shaker grate, insulated jacket, and poker. Optional battery-operated circulating pump will keep system operating when electric power is off.

JENSEN. *(wood/coal)* U.S. A steel plate boiler with cast iron heat baffles, firebrick combustion chamber, shaker grate and ash pan. A 37-cubic-feet-per-minute blower provides the combustion draft air. Domestic hot water tank included.

MENOMINEE. *(wood)* U.S. Wet leg designed steel plate boiler with firebrick-lined firebox. Large fuel door, V-bar grate, ash door, bypass damper, safety controls, and detailed instruction manual. Primary air supplied by thermostatically controlled blower. Two other models are MBW 6000, wood fired for smaller homes, and MBW 9000, for medium-sized coal boiler.

MONARCH. *(wood/coal)* U.S. A heavy-duty cast iron watertube boiler with heat exchanger suspended above combustion chamber. Cast iron-lined firebox, shaker grates, ash pan, large fuel door, electric damper control, safety controls, smoke shield, and insulated cabinet. Firebrick liner on coal model.

NECA. *(wood/coal)* Italy. A cast iron sectional boiler utilizing water cooled grates. Preassembled sections, manual stovepipe collar damper, non-electric aquastat-controlled primary damper, ash pan, insulated jacket, and cleaning tools. 10-year warranty.

ONEIDA ROYAL. *(coal/wood)* U.S. Steel plate, wet leg boiler with rocking type shaker grates. Equipped with automatic draft control, overheat control, hand flue damper, barometric damper, and insulated jacket. 5-year warranty.

RUSSELL. *(wood/coal)* U.S. A horizontal water tube steel boiler with afterburner to obtain complete combustion. Manual draft control, safety relief valve, slide bar tube cleaner and insulated jacket.

SEVERANCE. *(coal/wood)* U.S. Utilizes a coil pipe heat exchanger above an 18″ refractory-lined combustion chamber. Shaker grate, thermostatically controlled primary air, jacket, and top-loading fuel door.

SEVILLE. *(wood/coal)* U.S. Heavy steel wet leg designed boiler with firebrick-lined combustion chamber. Interior adjustable damper, bypass baffle, ash pan, jacket, and thermostatically controlled primary air blower. 5-year warranty.

TARM. *(coal/wood)* Denmark. Horizontal watertube heat exchanger above a round firebox. Cast iron shaker grates, automatic draft regulator, secondary air inlet, and enameled jacket. Firebox will hold large amount of coal. Model 202 is 80 pounds, Model 303 is 135 pounds.

THERMO-CONTROL. *(wood)* U.S. A steel plate boiler with firebrick-lined firebox and three internal serpentine pipe coils. Large airtight gasketed door, preheated primary and secondary air supply, and automatic thermostat on unit. No jacket. Will heat the room in which it is located.

TIMBER EZE. *(wood/coal)* U.S. Steel plate construction boiler with finned water reservoir. Firebrick-lined firebox, baffle system, shaker grates, insulated jacket, dual draft controlled by hydronic draft regulator, and ash pan.

UMCO. *(wood/coal)* U.S. Water backed steel plate boiler with cast iron shaker grate. Large loading door, ash pan, insulated jacket, thermostatically controlled primary air blower, cleaning brush, and fusible plug overheat control.

VALLEY FORGE. *(coal/wood)* U.S. A wet wall, steel plate boiler with heavy-duty shaker grate and firebrick-lined firebox. Non-electric aquastat controlled primary draft, large ash pan, firebox baffle, and optional jacket. Comes with complete safety controls and circulating pump.

MULTIFUEL BOILERS

The following are the specifications for some of the multifuel boilers manufactured. These will burn solid fuels and one of the conventional fuels, oil, gas or electricity.

Manufacturer and Model	Dimensions Height by width by depth	Water Capacity gallons	Weight pounds	Heating Capacity Btu/hour
Conseco				
HDG 25	50.8 × 16.6 × 42.3	22.9	1078	85,950
HDG 35	51 × 27 × 44.5	28.2	1310	120,000
HDG 50	51 × 27 × 46.2	40.4	1527	172,000
Dual Pak				
Model 35	48 × 28 × 55.3	36	1200	125,000
Dumont				
DWB 130	24 × 52 × 54.5	na	na	130,000
Essex				
	62.5 × 36.8 × 43.3	85	1180	175,000
Hogfors				
15 Maxi	52 × 33 × 24	51	na	40,000/ 50,000
15 Mini	38 × 33 × 24	21	na	40,000/ 50,000
Hoval				
VarioLyt 22	43 × 25.6 × 26.9	16	650	52,000/ 60,000
VarioLyt 30	43 × 25.6 × 28.9	18.5	675	64,000/ 84,000
VarioLyt 40	43 × 25.6 × 32.9	21	800	84,000/ 112,000
ZK 25	66 × 33.5 × 21.5	34	na	68,000/ 96,000
ZK 35	70 × 33.5 × 28	50	na	80,000/ 116,000
Neca				
SC 25	40 × 20 × 16.8	7.2	501	72,000
SC 35	40 × 20 × 20.5	8.7	587	108,000
SC 41	40 × 20 × 24.2	10.2	673	128,000
SC 47	40 × 20 × 27.9	11.7	759	152,000
New England Boiler				
WCO 120	50.5 × 25 × 50	60	1535	120,000
WCO 150	50.5 × 25 × 58	79	1881	165,000
Newmac				
O 95	48 × 24 × 52	65	890	160,000

Manufacturer and Model	Dimensions Height by width by depth	Water Capacity gallons	Weight pounds	Heating Capacity Btu/hour
Northland Boiler				
DF 620	47.3 × 21 × 42.5	63.5	1259	110,000
DF 790	47.3 × 21 × 50	73	1403	147,000
Oneida Royal				
CWOB 112	49.5 × 21 × 38.5	60	1259	112,000
CWOB 150	49.5 × 21 × 46	70	1403	150,000
Simplex Multitherm				
OCF 120 oil	52.1 × 22.5 × 25.5	na	1040	120,000
GCF 120 gas	52.1 × 22.5 × 25.5	na	1040	120,000
Stadler				
KL 21	59.6 × 36.5 × 22.1	50	990	68,000
KL 28	63.9 × 37.6 × 24.1	57	1122	90,000
KL 35	64.2 × 41.2 × 24.1	69	1210	112,000
KL 45	68.3 × 44.7 × 26.1	83	1430	144,000
UL 25	38.8 × 22 × 29.5	13	810	80,000
UL 30	38.8 × 22 × 32.3	16	832	96,000
UL 35	38.8 × 22 × 32.7	18.5	854	112,000
UL 40	40.2 × 24.4 × 34.6	22.5	975	128,000
Tarm				
OT 28S	51 × 35.8 × 24.8	73.5	946	72,000
OT 35S	51 × 39.5 × 30	76	1089	112,000
OT 50S	51 × 46.8 × 30	91	1444	140,000
Weber				
Mark V	50 × 23.5 × 25	8	700	80,000
Mark IV	50 × 23.5 × 34	10.75	900	100,000

CONSECO. *(wood/coal/pellets/chips/sawdust/oil/gas)* U.S. A single chamber magazine feed boiler that operates on the principle that the flue gases and particulates are recirculated through the fuel bed. Shaker grates, two ash pans, firebrick combustion chamber, combustion air damper, aquastat, and jacket with insulation. Features control panel and automatic ignition of solid fuel.

DUAL PAK. *(wood/coal/oil)* Canada, A two-pass firetube design boiler built from 5/16″ boiler plate. Firebrick-lined firebox, shaker grate, air-cooled fire door, insulated jacket, and flue brush. Alternate fuel in separate firebox is automatically switched. Includes delay timer and damper interlock with the flame retention oil burner.

DUMONT. *(wood/oil)* U.S. Designed to operate with an auxiliary heat storage tank, this unit achieves high efficiency. Water jacket, refractory brick firebox, draft inducer-blower and secondary burn chamber with flame retention burner. Also available as wood-only model.

ESSEX. *(wood/oil/gas/electric)* U.S. A gasifying down draft boiler that has a double-pass refractory combustion chamber. Features large wood chamber, oil burner located in secondary chamber, damper motor, and on-off mode of operation similar to oil or gas boilers.

HOGFORS. *(wood/coal/oil)* Finland. A dual chamber, multifuel cast iron boiler. Cast iron grates, large feed and ash doors with bimetallic aquastat-operated draft control, insulated jacket, and safety controls.

HOVAL. VarioLyt *(wood/coal/oil)* U.S. A solid-fuel boiler with an optional swing-away burner. Change from solid fuel to oil takes only a few seconds. Large firebox, control package, insulated jacket, and cleaning brush.
Series ZK *(wood/coal/oil/gas/electric)* Liechtenstein. A double-pass, bottom-burning combustion system with separate fireboxes. Large water jacket with bimetallic aquastat that controls damper on ash door. High efficiency conventional burner.

NEGA. *(wood/coal/oil/gas)* Italy. Similiar in design to LC series of add-on boilers. Switch-over accomplished by swinging open the solid fuel door and swinging in the panel-mounted oil or gas burner. Features a microswitch to cut off power to conventional burner if fuel door is opened.

NEW ENGLAND BOILER. *(wood/coal/oil/gas)* U.S. A wet leg, heavy steel plate boiler with separate combustion chambers. Cast iron shaker grates, safety controls, expansion tank, non-electric aquastat primary air damper, insulated jacket, and cleaning brushes for horizontal firetubes. Also available in steam units.

NEWMAC. *(wood/coal/oil/gas)* U.S. Similar to Newmac wood or coal boilers but with added flame retention head burner in separate combustion chamber. Available with several sized burners.

NORTHLAND BOILER. *(wood/coal/oil)* U.S. A horizontal firetube, wet leg boiler with separate oil burner chamber. Cast iron grates, ash pan, expansion tank, large fire door, enameled insulated jacket, and automatic wood fire draft regulator. 5-year guarantee.

ONEIDA ROYAL. *(wood/coal/oil/gas)* U.S. Heavy steel plate, wet leg boiler with horizontal firetube construction. Large feed door, shaker grates, insulated jacket, built-in tankless coil, automatic fire draft regulator, and cleaning brushes. Oil or gas burner in separate bolt-on chamber.

SIMPLES MULTITHERM. *(wood/coal/oil/gas)* U.S. A three-pass, pitched horizontal tube boiler with wet leg construction. Separate oil or gas combustion chamber, shaker grates, automatic damper control, insulated jacket, and door switch to shut off burner.

STADLER. KL Series *(wood/coal/oil/gas)* West Germany. Wet base boiler with separate combustion chambers for high efficiency. Ash pit, primary and secondary air supply, insulated jacket, and safety coil for overheat protection. Features a 32-to 58-gallon domestic hot water tank.

UL Series *(wood/coal/oil/gas)* West Germany. A wet base, water cooled fire grate boiler. Oil or gas unit on hinged door swings into ash door and only takes a few seconds to change. Features domestic hot water tank holding 40 to 53 gallons.

TARM. *(wood/oil/gas/electric)* Denmark. Designed with side-by-side separate fireboxes, wet base water jacket, and fixed grate. Tankless water heater, non-electric aquastat, primary and secondary draft control, cleaning tools, and insulated outer jacket. Optional shaker grates available for coal.

WEBER. *(coal/wood/oil)* U.S. Steel plate, vertical tube boiler with firebrick-lined firebox, cast iron shaker grates, and ash pan. Electronic automatic draft control and flame retention burner. Primary draft may be piped to use outside air. 10-year warranty.

STOKERS

The following is a list of some domestic sized furnaces and boilers with fuel stokers.

Manufacturer and Model	Dimensions Height by width by depth	Blower or Water Capacity	Weight pounds	Heating Capacity Btu/hour
Anthratherm				
Furnace				
VA 150H	53 × 36 × 62	2280 cfm	950	150,000
Anthratube				
Boiler 130-M	42 × 24 × 44.3	na	1000	130,000
Boiler 260-M	46.3 × 28 × 58.3	na	1530	260,000
Eshland				
Boiler Coal Gun	42 × 22 × 48	27 gal.	1000	130,000
Combustioneer				
Furnace				
24FA/S20	57.3 × 42.3 × 61.2	na	725	156,000
Dakota				
Furnace D280	57.5 × 29.8 × 67	1400/2600 cfm	910	96,000/ 350,000
Furnace C380	57.5 × 29.8 × 67	2000/4000 cfm	940	96,000/ 475,000

Manufacturer and Model	Dimensions height by width by depth	Blower or Water Capacity	Weight pounds	Heating Capacity Btu/hour
EMF				
Boiler DF520W	51.4 × 42.5 × 42.6	na	1250	145,000
Hardin				
Furnace FA 130	48 × 27 × 62	3140 cfm	700	130,000
Furnace FA 250	48.5 × 28 × 67	3100 cfm	950	250,000
Boiler FC-A-PK450	63.9 × 22 × 31.4	na	1450	144,000
Boiler FC-A-PK790	63.9 × 22 × 44.4	na	2200	252,000
Prill				
Furnace 82A	42 × 23.3 × 45.3	800 cfm	510	80,000
Furnace 100H	49 × 37 × 51	1200 cfm	830	154,000
Furnace 200H	57 × 44.5 × 62	2000 cfm	980	252,000
Stokermatic				
Furnace Mark II	38.5 × 28 × 39.5	1380 cfm	na	73,000
UMCO				
Boiler Mini Stoker	40 × 18 × 34	na	275	80,000

ANTHRATUBE. *(anthracite coal)* U.S. A wet leg steel boiler with overfeed stoker. Stepped reciprocating grate, helicoidal ribbon coal tube, induced draft fan, ash tub, centrifugal heat absorber, controls, and jacket. Takes coal from bin. Also adaptable to steam.

ANTHRATUBE. *(anthracite coal)* U.S. A wet leg steel boiler with overfeed stoker. Stepped reprocating grate, helicoidal ribbon coal tube, induced draft fan, ash tub, centrifugal heat absorber, controls, and jacket. Takes coal from bin. Also adaptable to steam.

ESHLAND. *(coal)* U.S. An overfeed stoker boiler with swirl chamber heat exchanger and cyclone ash separator. Utilizes a hollow center auger feed from bin or hopper, forced draft fan, automatic grate, insulated jacket, and controls. Available in sizes to 500,000 Btu/hour output for industrial applications.

COMBUSTIONEER. *(coal)* U.S. Large steel doughnut-type heat exchanger, firebrick-lined combustion chamber, and underfeed stoker. Firing rate to 20 pounds per hour with bin type stoker. Belt-driven blower, large plenums, filter, automatic controls, and jacket.

DAKOTA. *(coal/pellets)* U.S. Features a convoluted steel heat exchanger, underfeed stoker, and ash removal auger. Includes baffled firebox, combustion air blower, heated air blower, filter, controls, and jacket.

EMF. *(coal/oil)* U.S. An underfeed stoker system that contains a flame retention head oil burner backup. Steel plate construction, insulated jacket, ash tub, and hot water coil. Features ratchet drive stoker with capacity to 20 pounds per hour. Also available as a steam unit.

HARDIN. Furnace — *(coal/pellets)* U.S. Manufactured with aluminized steel heat exchanger and automatic stoker. Feed rate 10.4 pounds per hour for FA 130 and 20 pounds per hour for FA 250. Has large plenums and filter. Thermostatically controlled operation.
Boiler — *(coal)* U.S. A water leg design boiler with horizontal firetubes and underfeed stoker. Large ash pit, feed and cleanout doors, jacket, and controls. Also available as a steam boiler.

PRILL. *(pellets/coal)* U.S. A steel plate, two-pass furnace with underfeed stoker. Available as bin or hopper feed. Contains large ash pan, automatic controls, large warm air and return air plenums, filter, belt-driven blower, and jacket.

STOKERMATIC. *(coal)* U.S. Features a finned alloy steel heat exchanger and underfeed stoker. Integral 225-pound sealed hopper, floating feed screw, overfire air jet, ash bucket, large heat and cold air plenums, filter, electric controls, and jacket.

UMCO. *(pea or buckwheat anthracite)* U.S. An overfeed stoker boiler with 75-pound capacity hopper and automatic shaker grate. Forced fan draft system, ash pan, controls, insulated steel jacket, and installation/operation manual.

INDEX